THE SCIENCE OF
HARRY POTTER

THE SCIENCE OF
HARRY POTTER

THE SPELLBINDING SCIENCE BEHIND THE MAGIC, GADGETS, POTIONS, AND MORE!

MARK BRAKE WITH JON CHASE

Racehorse Publishing

Racehorse Publishing books may be purchased in bulk at special discounts for sales promotion, corporate gifts, fund-raising, or educational purposes. Special editions can also be created to specifications. For details, contact the Special Sales Department, Skyhorse Publishing, 307 West 36th Street, 11th Floor, New York, NY 10018 or info@skyhorsepublishing.com.

Racehorse Publishing™ is a pending trademark of Skyhorse Publishing, Inc.®, a Delaware corporation.

Visit our website at www.skyhorsepublishing.com.

10 9 8 7 6 5 4 3 2

Library of Congress Cataloging-in-Publication Data is available on file.

Print ISBN: 978-1-6315-8237-0
Ebook ISBN: 978-1-6315-8238-7

Printed in the United States of America

This book is dedicated to our daughters: Frances, Been, and Eden

TABLE OF CONTENTS

Part I—Magical Philosophy

Part II—Technical Trickery and Paraphernalia

Part III—Herbology, Zoology, and Potions

Part IV—Magical Miscellany

PART I
MAGICAL PHILOSOPHY

WHAT LIES BEHIND THE STUDY OF ASTRONOMY AT HOGWARTS?

Astronomy plays a dramatic but subtle part in the storylines of the *Harry Potter* series. It's against the backdrop of a full moon, of course, that we first see Remus Lupin, also known as Moony, transform from half-blood wizard into a werewolf in the *Prisoner of Azkaban*. It's the lunar light that triggers Lupin's lycanthropy.

The enchanted ceiling of the great hall at Hogwarts has been known to conjure up astronomy at night. A reflection of the sky above, the ceiling seems to zoom in on star clouds and swirling galaxies, as if trying to better Hubble.

And the astronomy tower, the tallest tower at the castle, is the setting for one of the series' most dramatic scenes. Under the gathering darkness of the death eaters' dark mark, lurking high above the tower, Dumbledore meets his death from a killing curse, cast by Severus Snape. But the astronomy tower is also where the students study. At midnight, under the stewardship of Professor Aurora Sinistra, they gaze at the planets and stars through their telescopes. So, what use is astronomy to wizards and witches in the Hogwarts curriculum?

Moons and Planets

Knowledge of the phases of the moon might come in handy. As werewolves transform under a full moon, knowing when the phases occur, no matter where in the world you are, would be useful for a wizard wishing to avoid lycanthropes. As for the planets, they define the very days of the wizarding week. In Latin, they run Sunday to Saturday as follows: *Solis* (sun/Sunday), *Lunae* (moon/Monday), *Martis* (Mars/Tuesday), *Mercurii* (Mercury/Wednesday), *Iovis* (Jupiter/Thursday), *Veneris* (Venus/Friday), and *Saturni* (Saturn/Saturday). As you can probably see, even in English, some of the planetary days remain: Sunday, Monday, and Saturday, still bearing the mark of sun, moon, and Saturn respectively in their names.

It seems the Hogwarts curriculum also required its students to learn and understand the movements of the planets. Such a study of the planets is not without a peculiar brand of British humor. Witness an encounter with the outer solar system in which Professor Trelawney, peering down at a chart, declares to Lavender Brown, "It is Uranus, my dear," only to hear Ron reply, "Can I have a look at Uranus, too, Lavender?" And Hermione's correction of Harry's understanding of Jupiter's moon, Europa, ". . . I think you must have misheard Professor Sinistra, Europa's covered in ice, not mice."

But much can be learned from even the merest mention of detail in the stories. Take, for example, the fleeting allusion in the *Harry Potter and the Sorcerer's Stone* scene where Hermione tests a reluctant Ron on astronomy, while Harry pulls a map of Jupiter toward him and begins to learn the names of its moons. In *Harry Potter and the Order of the Phoenix*, all three deal with a difficult essay on Jupiter's moons.

Tipping Point Cosmologies

The history of astronomy, like magic, is a long one. And for much of that history, the focus was the movement of the planets. One system, the geocentric system, puts the Earth at the center of the ancient universe. The planets speed in circular orbits about the central Earth. This scheme gives a good account of the sun's behavior on its yearly journey through the zodiac and the seeming path of the sun across the sky. The geocentric system also gives a reasonable account of the rather less regular motion of the moon. But the plain circular orbits come nowhere near explaining the observed motions of the wandering planets.

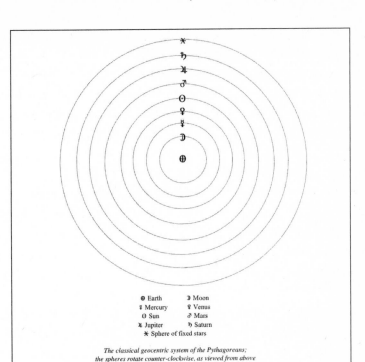

⊕ Earth   ☽ Moon
☿ Mercury   ♀ Venus
☉ Sun   ♂ Mars
♃ Jupiter   ♄ Saturn
✳ Sphere of fixed stars

The classical geocentric system of the Pythagoreans;
the spheres rotate counter-clockwise, as viewed from above

The geocentric system of Aristotle and Ptolemy

Pitted against the geocentric planetary system is the sun-centered, heliocentric cosmology. Here, the sun and its attendant planets are set out in their true heavenly order. The heliocentric system also explains the curiosity of the apparent movement of the planets. Such movement can only be understood when you know that the Earth itself is a moving planet. In the geocentric system, Earth is no mere planet, but the center of the entire universe.

The movement of the planets was the tipping point for your cosmology of choice. Both systems have been known since ancient times, but the mercurial events of the medieval period pushed an obscure Polish cleric, Nicolas Copernicus, to re-launch the heliocentric system in a book that would change history. Copernicus's book, *De Revolutionibus Orbium Coelestium (On the Revolutions of the Heavenly Spheres)*, was published in 1543. Its appearance meant the eventual defeat of the ancient and long-accepted geocentric system—but not without controversy.

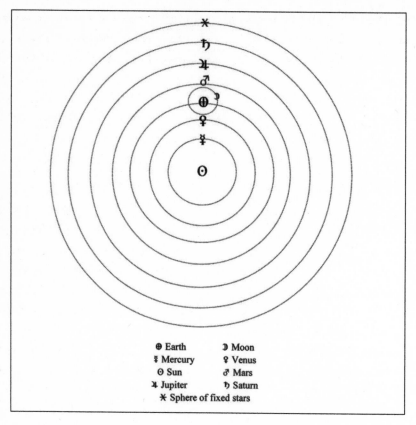

The heliocentric system of Copernicus

The Darkness Rising

The medieval church had supported the geocentric system. It placed man midway, between the inert clay of the Earth's core and the divine spirit. Man could either follow his base nature down to Hell, at the center of the Earth, or follow his soul and spirit, up through the celestial spheres to the heavens, so the system of planets became tied up with the medieval drama of Christian life and death.

To shift the Earth was to move the throne of God himself, who was meant to reside beyond the sphere of the fixed stars—and yet shift the Earth is exactly what Copernicus did. His new planetary system, and the theories it inspired about a boundless new universe teeming with planets, gravely disturbed Western philosophy and religion. Heliocentrism demoted the Earth from the center of the universe. It raised doubts about articles of Christian faith, such as the doctrine of salvation and the conviction of divine power over all earthly matters. And it called into question the nature of creation, and its relation to the creator. In short, Copernicanism raised doubts that were of fundamental importance to human, albeit muggle, identity.

The World Turned Upside Down

Then, there was Galileo. Italian mathematician Galileo Galilei wielded the newly invented telescope like a weapon of discovery and the new universe was unveiled. Galileo found Earth-like mountains and craters on an imperfect moon. Impure spots on the sun. Countless stars, worthy of adorning Dumbledore's cloak, which could only be seen with the aid of the spyglass. So much for the perfection and immutability of the heavens.

Galileo's most shattering discovery was the main four moons of Jupiter. They were proof of a focus of gravitation other than Earth. The old system had it that only the Earth acted as such a center. And when Galileo invited the great and good to peer at the moons through the spyglass, not one among the eminent guests was convinced of their existence. Some were so blinded by prejudice, they even refused to look down the tube. Galileo's optic tube shattered the old universe.

That battle was won with Galileo's discovery of these new worlds. It sparked the Scientific Revolution. It marked the paradigm shift of the old universe into the new. The cozy old geocentric cosmos was about man. The new universe of Copernicus and Galileo was decentralized, dark, and infinite. This is what sits behind the study of Jupiter's moons in the Hogwarts curriculum, and this branch of astronomy places the wizards in the progressive camp in the battle of the cosmologies.

DID NATURE, LIKE MAGIC, CONJURE SOMETHING OUT OF NOTHING?

There was the case of the three brothers, who conjured a bridge to get across a river. And another case, where a professor conjured a tea tray when he fancied a quick cup of tea. Or the young lady who, on separate occasions, conjured a flock of canaries to keep herself company and conjured a crystal flask to hold another professor's memory.

Conjuration was to transfiguration what cosmology is to physics—the tricky bit. Transfiguration was that branch of magic whose aim was to change the form or appearance of an object by altering an object's molecular structure in many cases. But Conjuration was the skill of transfiguring an object from thin air—from the very ether itself. This made Conjuration some of the most complex magic taught at Hogwarts, mostly to sixth year students and above. And there were limits to what could be conjured. According to *Gamp's Law of Elemental Transfiguration*, a law that governed the world of magic, there are some things that simply can't be conjured out of nothing, one of those being food. Birds and snakes, on the other hand, are a piece of cake. More than any other creature, these two are the easiest to conjure.

But snapping your fingers, or waving your wand, can sometimes spell danger. Some Conjurations could go wrong. This was especially true when conjuring creatures. If the Conjuration does not get carried out to the letter, or the caster was simply being rather silly with their skills, dangerous mistakes such as frog-rabbit hybrids could occur. Apparently, such monstrosities could be explained by the principle of artificianimate quasi-dominance: a kind of elemental superposition of cellular recon-struction. Severed heads and indeterminate stumps were another possible consequence. But can nature itself also conjure something out of thin air? And why, in the cosmos, is there something, rather than nothing?

The Beginning of Space and Time

We live in an evolving cosmos, and few things are evolving faster than our understanding of it, it seems. The universe of our ancestors was small, static, and geocentric. Now, in the 21st century, we find ourselves adrift in an expanding universe so large that light from its outer reaches takes longer than twice the age of the Earth to reach our telescopes.

Today, cosmology is almost universally conducted within the confines of the big bang theory. And this theory holds that the cosmos began in a 'singularity'—a state of infinite curvature of space-time itself. Now, in this singularity, all times and places were one. So, the big bang didn't happen in an already established space. Space was enmeshed *in* the big bang. The same goes for place. The big bang didn't happen in a specific place. It happened precisely where you are, and all other places at the same time. And finally, the big bang wasn't an explosion in a pre-existing space. Not in the way we normally think about explosions. Stuff didn't kaboom out into space, but stayed where it was, as the surrounding space expanded.

The evidence for this rather fantastic sounding big bang rests mainly on three main interpretations of physical evidence. Firstly, the redshift of whirling galaxies in space-time suggests the cosmos may be expanding. Run time backwards, and the expansion backs up into the singularity we spoke of above. Secondly, as the cosmos expanded it is believed to have left behind an afterglow, the Cosmic Microwave Background, a remnant radiation of the hot origin of the universe. And thirdly, the mixture of the very elements of the universe is taken as evidence. As the cosmos cooled down, the concoction of chemical elements began to be created, and evolved to give today's observed proportions. Lots of hydrogen, a fair amount of helium, and hardly anything else—that kind of thing.

But what happened before the big bang? And was the entire universe seemingly conjured out of this singularity? Cosmologists claim so.

How a Cosmos Might Have Been Simply Conjured

Now, most sensible people know there's no such thing as a free lunch. Most people, that is, apart from cosmologists. Many of the physicists whose job it is to explain the growing complexities of the big bang believe that the universe arose out of nothing. Nature, somehow, conjured up the cosmos.

When the scholars do the sums, they claim that our universe started out as a speck, which tunneled through an energy barrier to a larger radius, inflating into an expanding cosmos. The rest, as they say, is history. But this isn't all they claim. The scholars also suggest that their mathematical models that support the tunneling hypothesis don't vanish, even when the initial size of the universe is *zero*. In short, the cosmos could tunnel to some radius that permitted an inflation and expansion from *literally nothing*.

A word is needed here on the concept of nothing. By 'nothing', we don't mean the vacuum of empty space. This physical vacuum is rich with energy, particles, and antiparticles, which forever appear and disappear within it. The vacuum of empty space is not just a neutral theater in which things happen. And Einstein also held that space can warp and stretch. So, the 'nothing' of the big bang was a true nothing, a point beyond which space-time did not exist. And yet the cosmos was created.

Steady State and the Mini Big Bangs

Not everyone agrees with the big bang scenario. Starting in the 1940s, British physicist, Sir Fred Hoyle, and others worked on an alternative model of the universe that didn't start in an expansion. But this alternate theory, known as the Steady State Theory or C-Field theory, still believes in the creation of matter.

The Steady State holds that no big bang happened. Its proponents felt that the big bang's aesthetic association with thermonuclear weapons made the theory ugly and rather abrupt. And it also implied a creationist and mystical beginning to the cosmos.

In his 1994 autobiography, Hoyle wrote about what most annoyed him in the big bang theory. And that was its violation of the idea that the laws of physics are good for all corners of the cosmos, in all of space-time. This idea is held firm in the Steady State, but not in the big bang, which holds that at the very beginning of time, the warp of space was infinite, and the normal laws of physics broke down. Hoyle called this, "the crude breaking of the physical laws that occurs in big bang cosmology".

But the Steady State proponents still felt the universe could expand. And the creation of matter was key. Their theory was that the universe was infinitely old, and had no actual beginning. How could this be achieved? By the continuous creation of newborn matter, which compensated for the matter density lost by cosmic expansion. So, rather than having all matter created at the beginning of time, as in the big bang theory, the Steady State simply advocates a continuous creation of matter, a series of mini-big bangs, if you like.

Finally, why, in the cosmos, is there something, rather than nothing? Why did the universe, at least in the case of the big bang theory, simply appear? Because the laws of physics permitted it. In the theories of quantum physics, a process has a specific probability of occurring. No cause is necessary.

WHAT IS THE REAL STORY OF THE QUEST FOR THE SORCERER'S STONE?

I t was a legendary, rufescent stone, with magical powers. It helped create the Elixir of Life, which made its drinker immortal. And it transmuted any metal into pure gold. The Sorcerer's Stone has a special place in the Harry Potter Universe, created within the fiction by Nicolas Flamel, a real-life 14th and 15th century Parisian scribe and manuscript-seller. Harry's first battle against Lord Voldemort centered around the Stone during the 1991–1992 school year. Voldemort tried to steal the Stone for his own ends but was thwarted and his return to power postponed.

Once the Stone was secured, Dumbledore discussed its future with Flamel. The pair decided to destroy the Stone, with Flamel admitting he had enough Elixir left to set his affairs in order before he and his wife could contentedly die, after living for over 600 years. And yet, five years after the Stone's destruction, Harry wondered if as great a wizard as Voldemort might find another Stone. Perhaps the one created by Flamel was not unique. Besides, Voldemort was easily magically gifted enough to make his own. But what of the real-life quest to make a Philosopher's Stone?

The Magnum Opus

Alchemy is an ancient, and often secret, practice with roots the world over. Its study has occupied numerous philosophical beliefs, spanning thousands of years and countless different cultures. The persecution of many of its practitioners meant that alchemical traditions often adopted the habit of symbolic and cryptic language, which makes it hard to find links between the various alchemical cultures.

But, three main strands of alchemy can be identified. The alchemy of China and its sphere of cultural influence; the alchemy of India and the Indian subcontinent; and Western alchemy, which developed around the Mediterranean, and whose hub has shifted over the millennia from Greco-Roman Egypt, to the Islamic world, and lastly to medieval Europe. It may be that the three strands share a common ancestor, and have greatly influenced one another, but much variation in their traditions can also be seen. Western alchemy developed its own philosophical system, which shares some symbiosis with various Western religions.

Mention of the Stone itself can be found in writing from the beginning of the 4th century A.D. The Greek alchemist and Gnostic mystic, Zosimos of Panopolis, wrote one of the oldest known books on alchemy, called *Cheirokmeta*, Greek for "things made by hand."

Various recipes for the Stone exist, echoing the different cultures from which they came. Generally speaking, the recipe for the creation of the Stone followed an alchemical method known as *The Magnum Opus*, or *The Great Work*. Depending on the culture, the *Opus* describes work on creating the Stone, which passes through a sequence of color changes: *nigredo* (a blackening), *albedo* (a whitening), *citrinitas* (a yellowing), and *rubedo* (a reddening). The origin of the sequence can be traced to Zosimus and beyond. The *Opus* had a variety of alchemical emblems attached to it—birds such as the raven, swan, and phoenix were used to symbolize the sequence's progression through the colors. In practice, the alchemist would see the colors in the laboratory. For instance, *nigredo* could be seen as the blackness of rotting, burnt, or fermenting matter.

Base Metals to Gold

So, for many centuries, the Philosopher's Stone was the most coveted ambition in all of alchemy. The Stone had a long history. The ancient Greek atomist Empedocles was the creator of the cosmogenic theory of the four classical elements: earth, air, fire, and water. For Empedocles, these basic elements became the world of phenomena, full of contrasts and oppositions.

Through experiment he showed that invisible air was also a material substance, and proposed the order of the ancient elements as earth, water, air, and fire. Each element was above the other, tending if disturbed to return to its place in the natural order of things. Empedocles also held that opposite properties, such as love and hate, were material tendencies that mechanically mixed and divided in a continuous process. These ideas bear a similarity with the Yin and Yang dualism of ancient China. Though probably independent in origin, Chinese dualism also held that two principles, such as fire and water, male and female, forge to form other elements. In Chinese, these were metal, wood, earth, and through further fusion the 'ten thousand things' of the material world.

In this cosmogony, Empedocles forged a theory of everything. His worldview describes the separation of elements, the formation of ocean and earth, of moon and sun, and of atmosphere. Even further, he accounts for the biogenesis of plants and animals, and for the physiology of humans.

This elemental philosophy was also the alchemists' creed. Gold itself, along with the baser metals, such as mercury and lead, consisted of the elements of fire, air, water, and earth. So, it followed that, if you changed the proportions of those constituent elements, base metals could be transformed into gold. Gold was superior to the other metals as it was believed the very nature of gold meant it contained a perfect balance of all four elements.

Why Gold?

But why gold? Today, there are 86 known metals. But in ancient times, only seven metals were known: gold, silver, copper, iron, lead, tin, and mercury. We know them now as the Metals of Antiquity, and they would have been familiar to the ancient peoples of Mesopotamia, Egypt, Greece, and Rome. Of these seven metals, gold was the one that captured the human imagination and continued to do so for thousands of years.

Gold doesn't tarnish. It keeps its color. And it doesn't crumble. Gold seemed indestructible to ancient cultures. And yet it could also be easily worked. A single ounce of gold can be beaten to a thin sheet of gold metal 90 meters square.

Up until the year 1850, scarcely 10,000 tons of gold had been mined in all human history. One polar bear weighs around a ton, so it's the same weight of gold as 10,000 polar bears. That might sound like a lot, but this is all of human history we're talking about. Put another way, a blue whale weighs about 100 tons, so it's the same amount of mined gold as 100 blue whales, in all of history. You can see why someone might want to get their hands on more.

An Alchemist at Work

A good example of an alchemist at work is famous British physicist, Isaac Newton. In 16th and 17th century Europe, there were many who came to the courts claiming they possessed the secret to the Philosopher's Stone. So, throughout the continent, alchemists were employed by princes and nobles in the hunt for alchemical gold. For the alchemist, this situation was hugely profitable. A duke or a prince could be drained of substantial amounts of money during these hunts.

But alchemists didn't seek the Stone simply out of greed. Gold symbolized the highest state of matter. It personified human renewal and regeneration. A 'golden' person was resplendent with spiritual beauty, and would always triumph over the latent and lurking power of evil. The basest metal, lead, represented the sinful and unrepentant person, who was easily overpowered by the forces of darkness.

Like other alchemists, Newton searched ancient scripts for recipes. One such recipe, which Newton called 'The Net', was found in the writings of Ovid, the poet from the reign of the Roman emperor, Augustus. In his poem *The Metamorphosis*, Ovid tells the story of the god Vulcan finding his wife Venus in bed with the god Mars. According to the myth, Vulcan made a fine metallic net, captured the lovers within it, and hung them from the ceiling for all to see.

Now, in alchemy, Venus, Mars, and Vulcan stand for copper, iron, and fire. So, for the likes of Newton, the myth becomes an alchemical recipe.

And Newton indeed managed to synthesize a purple alloy, known as 'the Net', which was believed to be a step towards the Philosopher's Stone.

By recreating these recipes, modern scholars have found that Newton's alchemy included key elements of modern science—experiments that could be repeated and validated. And other members of the Royal Society were also alchemists. In many ways, alchemy was a secret practice of investigating the natural world.

THE DREAM OF ALCHEMY: WHERE DO BASE METALS TRULY TURN INTO GOLD?

In the Harry Potter Universe, alchemy is a branch of magic. It's an ancient science, which deals with the study of the four classical elements: earth, air, fire, and water. Magical alchemy also concerns itself with the transmutation of substances. So, it's linked to chemistry, potion-making, and the magic of transformation. Dating back to antiquity, alchemy is fused with philosophy, and mixed up with metaphysical and mystical conjecture. Even in the 20th century, there were still some members of wizard kind who actively studied magical alchemy. And, should there have been sufficient demand, alchemy was taught at Hogwarts, to those sixth and seventh year students who chose it.

Magical alchemy makes its presence felt in subtle ways within the Harry Potter Universe. Ancient alchemical texts often mention the chemical colors of red and white. Some scholars think that, like the base metals of silver and gold, red and white represented two different aspects of human nature. And the colors were the motivation for the forenames of Rubeus (red) Hagrid, and Albus (white) Dumbledore. In our own universe, the study of alchemy has run a parallel path to its magical counterpart.

The goals of alchemy were the creation of an elixir of immortality; the synthesis of an alkahest, or universal solvent, and chrysopoeia—the transmutation of base metals into noble ones, particularly gold. There is little doubt that alchemy, practiced throughout Europe, Egypt, and Asia, played a major role in the creation of early modern science, and especially medicine and chemistry.

But perhaps what's not so well known is the fact that contemporary scientists have realized the dream of chrysopoeia. For, somewhere in the unfeasibly huge cosmos, base metal elements *are* slowly being transformed into gold.

The Origin of the Classical Elements

Let's look at the backstory of the elements. The Ancient Greeks, among others, believed that the Earth was made up of the four elements: earth, air, fire, and water. Aristotle believed in a divine, but essentially dull and dormant cosmos. He imagined a two-tier, geocentric universe. The Earth, mutable and corruptible, was placed at the center. The sublunary sphere, essentially from the moon to the Earth, was subject to the transmutations of the four elements. This sphere alone was subject to the horrors of change, death, and decay.

Beyond the moon, in the supralunary sphere, the four earthly elements that catalyze change are gone. The rest of the cosmos, a nested system of crystalline celestial spheres, from the sublunary to the sphere of the fixed stars, is made of a different fabric—the quintessence, the fifth element. Chaste and immutable, the quintessence is flawlessly manifest in the shape of the crystal concentric orbs about the central Earth. And the further out we fly beyond the moon, the purer the quintessence becomes, until it meets its purest form in the sphere of Aristotle's God, the Prime Mover.

But Galileo's early experiments with the telescope put Aristotle's cosmos to the sword, for Galileo began to show that Heaven and Earth were made up of the same stuff. The moon was cratered and craggy, the sun's spots appeared to be an impurity in the quintessence, and the idea began to dawn in the minds of medieval scholars that matter was universal and mutable. In Galileo's words, *"What greater folly can be imagined than to call gems, silver, and gold noble, and earth and dirt base? ... These men who so extol incorruptibility, inalterability, and so on ... deserve to meet with a Medusa's head that would transform them into statues of diamond and jade, that so they might become more perfect than they are ... It is my opinion that the Earth is very noble and admirable by reason of the many and different alterations, mutations, and generations which incessantly occur in it."*

The Origin of the Chemical Elements

Fast-forward four centuries. By the late 19th century, the idea of the Periodic Table of chemical elements had begun to take shape. So, when the 20th century belief of the beginning and evolution of the universe—the so-called big bang theory—emerged, it had to take account of all things in the cosmos. And that included the origin and development of the chemical elements.

Early on, advocates of big bang cosmology realized that the universe is evolutionary. In the words of one famous cosmologist, George Gamov, "We conclude that the relative abundances of atomic species represent the most ancient archaeological document pertaining to the history of the universe." In other words, the periodic table is evidence of the evolution of matter, and atoms can testify to the history of the cosmos.

But early versions of big bang cosmology held that all the elements of the universe were fused in one fell swoop. As Gamov puts it, "These abundances . . ." meaning the ratio of the elements (heaps of hydrogen, hardly any gold—that kind of thing), ". . . must have been established during the earliest stages of expansion, when the temperature of the primordial matter was still sufficiently high to permit nuclear transformations to run through the entire range of chemical elements." It was a neat idea, but very wrong. Only hydrogen, helium, and a dash of lithium could have formed in the big bang. All of the elements heavier than lithium were made much later, by being fused in evolving and exploding stars.

How do we know this? Because at the same time some scholars were working on the big bang theory, others were trying to ditch the big bang altogether. Its association with thermonuclear devices made it seem hasty, and its implied mysterious origins tainted it with creationism. And so, a rival camp of cosmologists developed an alternate theory: the Steady State.

The Steady State held that the universe had always existed. And always will. Matter is created out of the vacuum of space itself. Steady State theorists, working against the big bang and its flaws, were obliged to wonder where in the cosmos the chemical elements might have been cooked up, if not in the first few minutes of the universe. Their answer: in the furnaces of the very stars themselves. They found a series of nuclear chain reactions at work in the stars. First, they discovered how fusion had made elements heavier than carbon. Then, they detailed eight fusion reactions through which stars convert light elements into heavy ones, to be recycled into space through stellar winds and supernovae.

And so, it's the inside of stars where the alchemist's dream comes true. Every gram of gold began billions of years ago, forged out of the inside of an exploding star in a supernova. The gold particles lost into space from the explosion mixed with rocks and dust to form part of the early Earth. They've been lying in wait ever since.

MERLIN: HOW DOES LEGEND COMPARE WITH CANON?

The most famous wizard in all of history—that's how Merlin is described in the Harry Potter Universe. And like many legendary people, his renown had led to his name becoming part of the everyday lexicon of wizard kind, such as the phrase, "Merlin's beard!", and the less common, "Merlin's pants!".

Canon has Merlin attend Hogwarts. Also known as the Prince of the Enchanters, Merlin is said to have attended the School of Witchcraft and Wizardry sometime during the medieval era. The omission of dates is wise, as legend is also suitably vague about arguably one of the most powerful myths of the English-speaking world. Canon rumor has it that his wand was made of old English oak, though his resting place has never been found, and so the truth of this has never been discovered.

Merlin was sorted into Slytherin House. Indeed, it is even said that Merlin was taught by Salazar Slytherin himself, one of the four founders of Hogwarts School. Famously, Salazar Slytherin's distrust for muggle-born students led to his leaving the school under a dark cloud of dispute. Merlin believed the exact opposite. He felt that wizards should help muggles and exist peacefully alongside them, echoing the representation of Merlin in T.H. White's wonderful *The Once and Future King*, where the wizard declares, "The Destiny of Man is to unite, not to divide. If you keep on dividing, you end up as a collection of monkeys throwing nuts at each other out of separate trees".

Such beliefs led to the Order of Merlin. Originally set up as a guard against the use of magic on muggles, the Order later morphed into an award, bestowed upon witches and wizards, for doing great deeds at personal risk, and for the betterment of wizard kind. This change in the emphasis of the use of the Order is perhaps a sign of the estrangement of the muggle world from the wizarding one. But who was Merlin?

Magicians of Medieval Science

Let's first put the legend of Merlin in context. For a thousand years, from the Fall of the Western Roman Empire to the Age of Discovery, very little science was done. The only research carried out was done so by clerics—priests, monks, and friars—for religious ends. And this is in great contrast to conditions at the time with Islamic science, where few scholars had religious leanings, and almost all experiments had utilitarian ends.

As a result of this, science became stagnant. The rise of Christianity in the West meant that from the 4th century on, intellectual life was articulated through Christian thought. Learning was confined to churchmen, and during the early Middle Ages, the history of thought over the lands of the disappearing Roman Empire was the history of Christian dogma. The Church fathers had set about their mission impossible, to integrate the more innocuous elements of the ancient wisdom of the Greeks into Christianity. Much of the old philosophy had already found its way in, by stealth. But the Old Testament and classical culture were unhappy bedfellows. And as the followers of Plato tried to crowbar in some safer aspects of philosophy, controversy was inevitable.

Heresy was born. Beginning in the fourth and fifth centuries, great disputes and heresies raged, particularly over Platonic ideas of the nature of the soul and its relation to corruptible bodies. Indeed, even the dispute over the nature of the Godhead itself was decided, as part of the great Arian heresy at the Council of Nicaea in 325 A.D.

By the 5th century, the deal was done. They had brokered a settlement between philosophy and faith, and science was in the doldrums. From now until the Renaissance, whenever questions were asked as to what people should believe in regard to religion, they were told it was not necessary to probe into the nature of things, as was done by those whom the Greeks call *physici*. Nor should pious people worry about being ignorant of the force and number of the elements, their motion, and order; eclipses of the heavenly bodies; the form of the heavens; the species and the natures of animals, plants, stones, fountains, rivers, mountains. Chronology and distances were not their concern, or the signs of coming storms. And the same was said over a thousand other things that philosophers had either found out, or thought they had found out.

It was enough for the everyday Christian to simply knuckle under. All they needed to do was believe that the only cause of all created things, whether heavenly or earthly, visible or invisible, was the goodness of the Creator, the one true God, and that nothing exists but Himself that does not derive its existence from Him. Talk about blunting the keen edge of curiosity and the intellect.

And yet there were men of the Middle Ages we could call philosophers. And Merlin certainly sounds like he would easily fit into such company. Scholars such as Roger Bacon (circa 1235–1315) sought to understand the ways of this world, through experience and reason. Some other philosophers of these times became prominent players in medieval society: Gerbert (circa 930–1003), one of the early scientists of the west, became a pope; Robert Grosseteste (circa 1168–1253), a very able philosopher, became a chancellor of Oxford University; and one of the most daring of thinkers during the late Middle Ages; Nicholas of Cusa (circa 1401–1464), became a bishop of Brixen. Any science these philosophers did was in their spare time.

The exceptions to the rule are Roger Bacon and the mysterious Peter the Pilgrim, a most Merlin-sounding man of the mind. Roger Bacon spent a considerable fortune on his researches in science. Despite a blessing from the Pope, Bacon was still thrown in prison for his pains. Peter the Pilgrim was a pioneer in the empirical study of magnetism and according to his great admirer Roger Bacon, "He does not care for speeches and battles of words but pursues the works of wisdom and finds peace in them."

Bacon predicted motor ships, cars, and airplanes. He also foresaw an alchemical science, "which teaches how to discover such things as are capable of prolonging human life". And yet so talented a man as Roger Bacon eulogized Peter the Pilgrim still further:

"He knows natural science by experiment, and medicaments and alchemy and all things in the heavens or beneath them, and he would be ashamed if any layman, or old woman or rustic, or soldier should know anything about the soil that he was ignorant of. Whence he is conversant with the casting of metals and the working of gold, silver, and other metals and all minerals; he knows all about soldiering and arms and hunting; he has examined agriculture and land surveying and farming; he has further considered old wives' magic and fortune-telling and the charms of them and of all magicians, and the tricks and illusions of jugglers. But as honor and reward would hinder him from the greatness of his experimental work he scorns them".

The Source of the Magicians

Medieval 'magicians' such as Bacon, Pilgrim, and Merlin, belonged to a long line of eminent philosophers. Little of the very early philosophers is known. But it appears that groups of sages, during the Iron Age, set up religious orders that doubled as philosophic schools. Those that flourished advised democratic chiefs or tyrants (at that time the word tyrant carried no ethical censure) giving advice on a range of policy and topics. To be sure, it bestowed kudos on a regime to have an eminent philosopher in tow.

So, the rise of Iron Age civilizations created a new social type in the form of these philosophers. The fact that knowledge of these early thinkers has survived, that Raphael painted a masterpiece in their honor, and that legends about their lives has lingered, shows just how significant they must have been in the ancient world.

The irresistible rise of the philosopher was a global phenomenon. The impact of the Iron Age was felt strongly in many parts of the developed world. In ancient China, Confucius and Lao-tze acted as political or technical advisors. In early India lived the rishis and Buddhas, Siddhārtha Gautama, the Buddha, being the most renowned. And in olden Palestine the prophets and the subsequent authors of the Wisdom literature, such as Ecclesiastes and the Book of Job, were alive and writing. Many of these thinkers and workers consorted with princes and tried in vain to reform their governments. But the important point is this: they all shared an interest in formulating a worldview of man and nature.

In some works of legend, Merlin serves as a king's advisor. As English poet Alfred Tennyson says of the wizard in his 1859 poem, *Merlin and Vivien*, ". . . the most famous man of all those times, Merlin, who knew the range of all their arts, had built the King his havens, ships, and halls, Was also Bard, and knew the starry heavens." And the original source of the myth of Merlin, Geoffrey of Monmouth, said in his *Vita Merlini* of 1152, "I knew the secrets of things and the flight of birds and the wandering motion of the stars and the gliding of the fishes . . . All this vexed me and denied a natural rest to my human mind."

Merlin the Magician

Merlin in legend is best known as the wizard in Arthurian myths, and medieval Welsh poetry. The primary historical account of the Merlin legend stems from Geoffrey of Monmouth's *Historia Regum Britanniae*, written around the year 1136. Geoffrey's work was based on an inventive amalgamation of earlier historical and legendary figures. He fused existing stories of Myrddin Wyllt (Merlinus Caledonensis), a North Brythonic philosopher and Wildman, with no link to King Arthur, with stories about the Romano-British warrior and rousing leader, Ambrosius Aurelianus. In so doing, Geoffrey created the composite figure he named Merlin Ambrosius or, in Welsh, Myrddin Emrys. Merlin is supposedly buried in the Broceliande forest, in Brittany.

Geoffrey's tales of the wizard were instantly popular, especially in Wales. After Geoffrey, later writers embellished his account to conjure a fuller image of the wizard. Merlin's biography casts him as a cambion: a mud-blood, born of a mortal woman, but sired by an incubus, a non-human from whom Merlin inherits his supernatural powers and abilities. The name of Merlin's mother is not always given, but is said to be Adhan, in the oldest version of the *Prose Brut*. In later, and creative Arthurian accounts, Merlin matures to an ascendant sagehood, and engineers the birth of Arthur through magic and intrigue. Later authors still place Merlin in the tradition we spoke of above—serving as king's advisor. Until, of course, he is bewitched and locked up by the Lady of the Lake.

The very name 'Merlin' is drawn from the Welsh *Myrddin*, the name of the bard Myrddin Wyllt, one of the main sources for the later legendary figure. In the same way that much is Latinized in the *Harry Potter* works, Geoffrey of Monmouth Latinized Merlin to Merlinus in his works. But the phrase 'Clas Myrddin', or 'Merlin's Enclosure', is thought to be an early name for Great Britain itself, as stated in the Third Series of Welsh Triads. Indeed, some Celticists believe there is a town named for Merlin, and that the Welsh name Myrddin was derived from the toponym Caerfyrddin, the Welsh name for the town known in English as Carmarthen.

The Merlin myths are so strong in British culture that Merlin is meant to have created Stonehenge. Geoffrey of Monmouth's account of Merlin Ambrosius' life in *Historia Regum Britanniae* is based on the story of Ambrosius in the *Historia Brittonum*. Geoffrey accounts for many of Merlin's prophecies, taken from his earlier work *Prophetiae Merlini*. And the most notable of these is the creation by Merlin of Stonehenge as a burial place for Aurelius Ambrosius.

So how do we place Merlin's legend in context, given what we've said? We talked of how Christianity eclipsed science in Europe for over a thousand years. Merlin's legend is meant to come from the very start of that period of a dark age for science in the West. And when Christianity prevailed, Celtic paganism and a closeness with nature migrated into mythology. Water and islands kept their magic. Wise old sages like Merlin knew of the old magic, the old ways of man and nature, the kaleidoscope of possibilities that was now shut away, blinded by the dogma of the Church. Merlin's legend conjured up ancient sophists of the Iron Age, his reputation sat alongside that of Peter the Pilgrim, and gave light to a hopeful future in a dark age of faith.

WHO REALLY WAS THE LAST GREAT WIZARD?

Who was the last great wizard? Professor Albus Dumbledore, perhaps? Famous for the discovery of the twelve uses of dragon's blood, and his work on alchemy. Or maybe, if you are more inclined to the dark and death eater side, you favor Tom Marvolo Riddle, later known as Lord Voldemort? The most powerful of dark wizards, Riddle claimed to have pushed the boundaries of magical forces farther than ever before.

In fact, the last great, and often dark, wizard was Isaac Newton. Newton's work had beauty, simplicity, and elegance. He is widely thought to have made the greatest work of science ever created. Newton was the 17th century British natural philosopher who first uncovered the laws of physics that govern the cosmos. He made up new branches of mathematics, conquered the composition of light, and divined the laws of gravity and motion, which hold sway across the entire universe. Newton ushered in an age, the Newtonian Age, based on the notion that all things in the cosmos were open to rational understanding.

But in 1936 a huge archive of Newton's private manuscripts was put up for auction at Sotheby's, in London. The papers had been kept from the public for over two centuries. One hundred lots of the manuscripts were bought by the famous British economist, John Maynard Keynes, who found that many of Newton's papers were written in a secret cypher. And for six years, Keynes struggled to decipher them. He hoped they would reveal the private thoughts of the founder of modern science. But what the code really revealed was another, far darker, side to Newton's work. For, in the manuscripts, Keynes found a Newton unknown to the rest of the world—a Newton obsessed with religion, and a purveyor of practices of heresy and the occult.

The Alchemist

The days of Newton were mercurial indeed. In Newton's time, England saw the Great Fire of London, the plague, and a Civil War that led to the death of 190,000 of his countrymen, out of a population of only five million. It was also a time that saw the dawning of the scientific revolution, an age when science and reason would redefine the world.

But the modern idea of Newton is about as far as could possibly be from what Newton himself thought. His private manuscripts show that, in the year he became a professor at Cambridge, Newton also bought two furnaces, a hotchpotch of chemicals, and a curious collection of books. Newton had found alchemy.

Alchemy at the time had been outlawed. In these desperate days of confusion and crisis, the government feared that frauds would corrupt the economy with bogus gold. And, if you got caught practicing alchemy, the punishment was swift and severe. Defrocked alchemists were habitually hanged on a gilded scaffold. And sometimes they were made to wear suits of tinsel as they were hanged, to make it even more of a public spectacle.

By the mid-1670s, Newton had shied away from the international stage of science. He swore never to publish another scientific paper. Instead, in the isolation of Cambridge, Newton threw himself into his alchemy. But Newton wasn't looking to make himself rich. He simply wanted to know the mind of God himself. As his Cambridge amanuensis, Humphrey Newton, said of Isaac, "What his aim might be, I was not able to penetrate into, but his pains, his diligence, at these set times made me think he aimed at something far beyond the reach of human art and industry." Orthodox scholars dismissed Newton's alchemy as worthless. But they missed the point. For Newton, alchemy was route one to God himself.

Alchemy was medieval matter theory. It was a science that sought answers to the most basic questions, such as, "What is the Earth? What is the cosmos made of? What are the constituents of matter?" Newton looked into the most ancient of texts in search of his answers. He believed that the Ancients understood great truths about nature and the cosmos. This wisdom had been lost over time, but Newton believed himself to be God's emissary on this Earth. He believed it was his task to find the secret cyphers, hidden in both the Bible and the Greek myths, which he interpreted as encoded alchemical recipes.

Newton's alchemy was an occult way to investigate natural philosophy. His practice of alchemy was early modern chemistry. Newton was experimenting with strange spirituous substances, and looking to transform materials from one form into another.

Gravity

Alchemy might have continued as Newton's obsession. But, in the 1680s, the astronomer Edmond Halley, now known for the comet, asked Newton what kind of curve would be described by the planets. Halley and others had begun to suspect that the planets were attracted to the sun by some strange kind of force. Halley's question would change the world forever. For the next eighteen months, Newton worked on the question of how the planets move through space. He barely ate, slept, and saw next to no one until finally, Newton created his 500-page masterpiece, the *Principia Mathematica*, considered by many to be the most magnificent, all-encompassing, and daring science book ever written. Newton had published a system of the world, a theory of everything. He saw that the moon's orbit about the Earth, the motion of the moons around Jupiter, and a cannonball's motion on Earth, all had a common cause of motion. They were governed by the same law of gravity. In a revolutionary leap, Newton declared this invisible force to operate everywhere in the cosmos. It was his universal law of gravitation.

And yet, Newton didn't understand the source of this gravity force. How can one object be attracted to another if there is nothing in between them? Some scholars believe that Newton's notion of gravity was related to the occult practice of alchemy. They believe that Newton's fascination with the vegetation of metals, in which inert metals seem to live and grow like plants, also applied to the mysterious and invisible force of gravity itself. It was action at a distance.

So, the last great wizard was not Dumbledore, or Voldemort, but Isaac Newton. Newton brilliantly conjured the tricks of his trade in every field in which he worked. He was an ingenious and energetic builder who dabbled in the dark stuff, as well as the light. He was astonishingly brilliant in great books like the *Principia* and in the genius of his daring experiments. As Newton's biographer, Richard Westfall, put it, "I have never, however, met one man against whom I was unwilling to measure myself, so that it seemed reasonable to say that I was half as able as the person in question, or a third, or a fourth, but in every case a finite fraction. The end result of my study of Newton has served to convince me that with him there is no measure. He has become for me wholly other, one of the tiny handful of supreme geniuses who have shaped the categories of human intellect, a man not finally reducible to the criteria by which we comprehend our fellow beings."

HOW WOULD HERMIONE'S TIME-TURNER WORK?

P icture this: Hermione is standing at the Crucifixion. Spellbound and open-mouthed, she can't help but stare at the scene. Perhaps the most famous in all of history. It was one of the benefits of time travel. Experiencing history unfolding firsthand. Just a few points to remember: she must do nothing to disrupt history. (Note to self: no stone throwing this time.) And when the crowd is asked who should be saved, she should join in with the call, "Give us Barabbas!" Suddenly, Hermione realizes something about the crowd. Not a single soul from 33 AD is present. The mob condemning Jesus to the cross is made up lock, stock, and smoking wand, of wizards from the future.

The entire scene is not just littered with wizards from the future. They've actually *changed* the outcome of history itself, by being present at the Crucifixion. The wizards think they know the way the story is meant to go. Rather than Jesus being set free, the crowd is meant to choose Barabbas, the bandit. But the decision only goes that way because the wizards are witness to the scene. Would Jesus have been set free instead, if they hadn't interfered? This would be exactly the kind of chaotic paradox that explains why The Ministry of Magic placed hundreds of laws around the most common wizard means of time travel: the time turner.

Time Turners

These time travel devices resembled an hourglass on a necklace. The number of times the hourglass was turned determined the number of hours a traveler could go back in time. Typical time turners, supplied by the Ministry of Magic, had an hour-reversal charm worked into them. This hour-reversal charm, encased within the device, was for added stability, and ensured that the longest period relived, without the prospect of serious harm to the traveler, was around five hours.

There also existed a "true" time turner that allowed the traveler to visit whatever time they liked, and far beyond the five-hour boundary. But few travelers ever survived such journeys. Trials with true time turners ended in 1899, when traveler Eloise Mintumble got trapped for five days in 1402 AD. Her body aged five centuries when it returned to the present and was fatally wounded.

Hermione was given a time turner by Professor McGonagall so that she could attend more classes than the Hogwarts timetable would allow. At the end of the school year, she and Harry also used it to travel in time to save Sirius Black and Buckbeak from certain death.

The prospect of the time travel paradox is common to most forms of time travel, and not just time turners. It's one of the reasons Professor Stephen Hawking refuses to believe such travel is possible. His argument goes something like this: "if time travel really *is* possible, then where are the time tourists of the future? Why aren't they visiting us, telling us all about the joys of time travel?"

Time Travel

Tampering with time has long been a sorcerer's dream. What if time could be mastered? What if this brutal agent, which devours beauty and life, could be tamed? The wizard world has four dimensions. Three dimensions are space; time is the fourth. There appears to be no difference between time and any of the three dimensions of space, except that our consciousness moves along it.

There have been folkloric flirtations with time, where dreamy magic is mixed with myth. And there has been the mechanized notion of time travel. Time travel devices are tied up with the concept of time itself. The ancient Greeks had two words for time, *chronos* and *kairos*. *Kairos* suggests a moment of time, in which something special happens. *Chronos* is more concerned with measured, sequential time. Natural philosophy brought a mechanistic approach to nature. *Chronos* came to the fore, and time travel devices were born.

Time was in the ether. It splashed upon the canvas of the Cubists. Artists such as Picasso and Braque produced paintings where various viewpoints were visible in the same plane, at the same time. All dimensions were used to give the subject a greater sense of depth. It was a revolutionary new way of looking at reality. Time was captured in cinema, and the stop-motion photography of Étienne-Jules Marey. It inspired Marcel Duchamp to paint his highly controversial *Nude Descending a Staircase*, which depicted time and motion by successive superimposed images. Americans were scandalized.

Spacetime was born. Einstein gave us a new perspective on the fourth dimension. Moving clocks run slow. Time is slowed down by gravity. And the speed of light is the same no matter how the observer is moving. It was a revolution in time. And it seemed to worry Salvador Dali. For many, his anxiety is palpable in his famous painting, *The Persistence of Memory*. The floppy clocks are history's most graphic illustration of Einsteinian gravity distorting time.

How Time Turners Tick

How would a time turner work? One possible way is through the creation of a wormhole. Famous American science fiction writer John Campbell was the man who invented such space warps. In his 1931 story, *Islands of Space*, Campbell used the idea as a shortcut from one region of space to another. And in his 1934 story, *The Mightiest Machine*, he called this same shortcut hyperspace, another now-familiar phrase.

A year later, world-famous Nobel Prize–winning scientist, Albert Einstein, with his colleague, Nathan Rosen, came up with the science behind the invention of time travel. They worked out the scientific theory that explained the notion of bridges in space. It was much later that scientists started calling these bridges wormholes.

Imagine we create a wormhole. A wormhole is a region of space that is warped. It's basically a shortcut in space and time through which to travel. The trouble is, though, wizard time travelers would not be able to travel back in time to a date before the wormhole was created. For example, if a wizard managed to create a wormhole on April 1, 1666, they wouldn't be able to go back in time before 1666. So, some splendid wizard in the distant past must have conjured up a wormhole to get the whole thing going.

So, what does a wormhole look like? It's the kind of swirly cosmic tunnel often depicted in movies when something is on a journey through space and time. A wormhole has at least two mouths, connected to a single throat. And scientists really *do* believe they exist. At least in theory. And, as that theory is Einstein's, people take it seriously. Stuff may travel from one mouth to the other by passing through the wormhole. We haven't found one yet, but the universe is immense. And we haven't really been looking very long.

DOES SCIENCE HAVE LIMITS, LIKE J.K. ROWLING'S LIMITS OF MAGIC?

Nature is its own magic. In the Harry Potter Universe, magic is portrayed as a supernatural force that, when used skillfully, can supersede the normal laws of nature. But, as the laws of nature are pretty well thought out, having stewed and simmered over thirteen-odd billion years of evolution, it's wise to wonder the way in which magic should be allowed to supersede, and in what situations. And, rumor has it, that's exactly what J.K. Rowling wondered about before publishing the first novel. For five years, the story goes, she established the limits of magic for her fantasy—deciding what magic should, and should not, be allowed to do. "The most important thing to decide when you're creating a fantasy world," Rowling said in 2000, "is what the characters can't do."

Hence, the Principal Exceptions to Gamp's Law of Elemental Transfiguration. Gamp's Law was a law governing the magical world. And food was the first of the five Principal Exceptions: witches or wizards could cook food with magic, but not conjure it out of nothing. When food did appear to have been conjured from nothing, such as McGonagall's self-refilling plate of sandwiches or the hordes of food during banqueting at Hogwarts, it was either being multiplied or transported from elsewhere.

Neither was the magical world full of get-rich-quick wizards. Rowling is on record as suggesting that, though not explicitly stated in the series, wizards could not simply conjure money out of thin air. An economic system based on such a possibility would be grimly flawed and highly inflationary. Perhaps that's also why a limit was placed on the use of the Sorcerer's Stone for alchemy. The Stone's abilities were described as extremely rare, possibly even unique, and possessed by an owner who did not exploit its powers.

Consider love and death. Some magical spells needed an emotional input while casting them. The Patronus charm needed the caster to focus on a happy memory. For example, Harry conjured a corporeal Patronus when Sirius was on the verge of receiving the Dementor's Kiss. Harry's force of will was an essential ingredient in the magic. Indeed, love is portrayed as a powerful form of magic. Love was, according to Dumbledore, a "force that is at once more wonderful and more terrible than death, than human intelligence, than forces of nature." In the *Goblet of Fire*, Dumbledore also says there is no spell that can bring people back from the dead. Sure, they can be re-animated into compliant Inferi on a living wizard's command. But they are little more than soulless zombies, with no will of their own. The limit is much referenced in the series, and wizards try transcending it at their peril.

Nor is it possible for a wizard to achieve immortality. Not without the Sorcerer's Stone or a horcrux—or seven. The three Deathly Hallows were fabled to bestow upon the owner the gift of being the master of death. But even then, it was hinted that a true master of death was really a wizard who was willing to bow to the inevitability of death.

And yet, death is still a fascination to wizard kind. In the Department of Mysteries sits a chamber containing an enigmatic veil. The veil is the divide between life and death. This manifestation of the barrier between the land of the living and the land of the dead is no doubt studied by the Unspeakables—witches and wizards who work at the Department. But does science also have limits?

Science, Fantasy, and the "What If?" Question

Fantasy is the faculty of imagining improbable things. But, then again, so too is science. Sometimes. Fantasy is a literary device for exploring imagined worlds, and in that sense is a kind of theoretical science. Scientists also make models of imagined worlds. They just happen to be more mathematical. They construct idealized universes and set about tweaking the parameters to see what might happen. The "what if?" question is key to both science and fantasy.

What if magic was real? What if a wizard could conjure up a thousand things? What if witches could wave wands and make them work? Scientists try to answer "what if?" questions, too. But they are bound to stay within the confines of the known theories of science. Fantasy writers have far more scope. And the very best fantasy can be used to consider profound philosophical or moral issues: the metaphysics of soul-splitting; the driving force of revenge; or deep questions about love and death. Have the big mysteries all been solved, and all the big questions answered? Is the age of the truly great discoveries behind us? And will there be a final "theory of everything" that marks the limits of science?

As early as 1968, Stanley Kubrick's *2001: A Space Odyssey* portrayed a scientific culture winding down. In the movie, space travel is replete with corporate logos and trademarks, showing a world absolutely stage managed. The corporate logos that appeared throughout the movie seemed, at the time, very sinister—a patent circumvention of democracy. The irony of the picture's portrayal of an insipid future dominated by corporations and technology was lost on some. Microsoft mogul Bill Gates suggested that *2001* inspired his vision of the potential of computers (though whether Gates was also inspired by the portrayal of sinister corporate domination is mere speculation). Nevertheless, such corporate control is symptomatic of the scientific crisis portrayed in the film.

Kubrick notwithstanding, there's an unwritten assumption that science is infinite and endless. But there also exists the idea that, one day, scientists may find such truths that would mean no further science will be necessary. There are even those who feel that this situation may soon be upon us, that we are already approaching the very limits of scientific knowledge: physicists who advance upon a theory of everything; evolutionary biologists who draw near a determination of how life on Earth began; cosmologists homing in on a theory of the creation of the cosmos; and neuroscientists probing a final understanding of consciousness.

Science also places limits on itself. Einstein's theory of Special Relativity places an upper limit on the speed of matter or information; quantum mechanics means that our investigation of the nanoscopic will remain indeterminate; chaos theory suggests many phenomena may be impossible to predict; and evolutionary biology reminds humans that they are mere animals, not relentless robots seeking the profound truths of nature.

Optimists may say these limits can be surmounted. And yet, many of the final questions may never be answered. We may never be able to probe the very beginning of the universe, if, indeed, there ever was one. We may never find if quarks and leptons are composed of smaller particles still. And we may never be able to fathom how inevitable life's origin was on Earth, or if there is life elsewhere in the cosmos.

And yet, maybe the answer is in the machine. Maybe sometime soon humans will create artificially intelligent machines, capable of totally transforming our limited science. In the most ambitious version of this scenario, the intelligent machines will be able to transform the entire cosmos into a gargantuan, holistic, data-processing network. Maybe then, when all matter becomes mind, we would be able to answer the ultimate question of why there is something rather than nothing. Now that *would* be like magic.

WHAT KIND OF PROPHECY
IS POSSIBLE?

"The one with the power to vanquish the Dark Lord approaches . . . Born to those who have thrice defied him, born as the seventh month dies . . . And the Dark Lord will mark him as his equal, but he will have power the Dark Lord knows not . . . And either must die at the hand of the other for neither can live while the other survives"—*Harry Potter and the Order of the Phoenix*

"If we have learned one thing from the history of invention and discovery, it is that, in the long run—and often in the short one—the most daring prophecies seem laughably conservative" —Arthur C. Clarke, *The Exploration of Space* (1954)

In the Harry Potter Universe, a prophecy was a prediction made by a Seer. The Seer, a gifted wizard or witch who was able to see into the future, would begin reciting the prophecy involuntarily, and would then go into a type of trance, while speaking strangely in a changed voice. The record of such a prophecy was then kept in a spun-glass orb and known, appropriately enough, as a prophecy record. The prophecy orbs, spherical objects that seemed to contain clouds of swirling mist, were kept in the Hall of Prophecies, housed within the Department of Mysteries. Only those members of wizard kind mentioned in the prophecy were allowed to remove that record from the Hall. Many prophecy records were destroyed in the Battle of the Department of Mysteries.

One of the most prominent Seers was Professor Sybill Trelawney. The Head of the Divination at Hogwarts School, Trelawney's first recorded prophecy was witnessed by Albus Dumbledore. The prophecy foretold the birth of a wizard who would be able, though by no means guaranteed, of defeating Lord Voldemort. The prophecy went on to say that Voldemort would mark this young wizard as his equal, and that either the young wizard, or Voldemort, would eventually kill the other. This boy, of course, was shown to be Harry Potter. Harry knew nothing of the prophecy until Dumbledore told him the tale, after the Battle of the Department of Mysteries. The prophecy proved to be very prescient. But what of prophecy in the muggle world? Is it possible? And, if not, what's the closest we can come to prophecy?

Muggle Prophecy

In the muggle world, prophecy belongs, if anywhere, to science. Science is different than other intellectual disciplines, such as the humanities, art, or religion. Science revolves around its practical application. Science is a discipline concerned with how things are done, and how outcomes can be predicted from practice.

Think of science as a recipe for doing things. Science shows you how to carry out certain tasks, should you need to do them, and what will happen when you do so. There's great power in this simple philosophy. It is fundamentally a philosophy of matter in motion, an account of nature and society from below rather than above. It is a philosophy that realizes the power of change through getting to know nature's ground rules.

Think also of the way in which science has evolved. History shows a distinct sequence of the emergence of its different disciplines. The order is usually: mathematics, astronomy, mechanics, physics, chemistry, and biology. The origin and development of this sequence lies in the concern with practical techniques to provide for human needs. This developmental time sequence of the sciences is fascinating. It appears to fit very well to the patterns of social advance. Notice how the sequence corresponds quite closely to the practical uses that were expected, if not demanded, of science by ruling classes at different times.

In ancient times, science derived from the techniques that arose from our concern with nature. For example, from the beginning of recorded history and the development of surplus, mathematics arose out of the need to make calculations relating to taxation and commerce, or to measure land. Observations of the sky were used to determine the seasons, an important factor in knowing when to plant crops, as well as in understanding the length of the year. These priestly functions gave rise to astronomy, of course.

Only much later did humans develop a fascination with the control of inanimate forces of physics. The demands of the new textile industry, the interest of the emerging manufacturers of the 18th century, gave rise to chemistry. The more complex sciences, such as medicine and biology, were developed through the study of the subject itself, with practically no input from the simpler sciences, like mechanics. Revolutionary discovery led the way in all these fields of science.

When we consider detailed examples of these sequences of discovery, other general trends tend to occur. In any specific discipline, a series of associated findings can be identified. The chain of events usually begins with an unexpected and revolutionary discovery, the coming together of fields previously thought unrelated, and ending up with an entirely new field of science. Hardly prophecy.

Consider the case of the Newtonian "System of the World". This was Isaac Newton's theory of everything, associated with the development of the theory of universal gravitation, in the late 17th century. The long chain of events leading to Newton's work began at least a century before with Copernicus' revolutionary proposal of the sun-centered planetary system. This led to the coming together of Galileo's experiments in terrestrial dynamics with Kepler's celestial mechanics, and ended up with Newton's synthesis of the new mechanical worldview of the universe, a worldview that was to dominate physics until the early 20th century.

Newton's worldview was that of a clockwork universe. Physics became regarded as the ultimate explanatory science: phenomena of any kind, it was believed, could be explained in terms of mechanics, and the cosmos was as a perfect machine, which was essentially open to prediction. The laws of mechanics, it was thought, could tell you exactly where Jupiter would be a week next Wednesday. And yet muggles soon discovered chaos is the law of nature, and order the dream of man.

Physics is fine if you wish to describe planets in orbit, spaceships sailing to Saturn, that kind of thing. But there are some aspects of nature that physics predicts badly. Turbulence is one such example. The air racing around a jet's wing. Blood coursing through the heart. Or even climate change. The behavior of complex systems like weather and climate is hard to model. Even if you could understand it, you still couldn't make precise predictions. Weather prediction is almost impossible. And that's because the system's behavior is acutely dependent on initial conditions, and tiny differences become hugely amplified.

Chaos is not just random and unpredictable. Because of the myriad tiny imperfections implicit to complex systems, tiny nuances soon start to make a difference. Soon enough the imperfections overpower your careful calculations, and even simple systems show unpredictable behavior. So, in fact, it turns out you can't really predict more than a few seconds into the future.

HOW DO VOLDEMORT'S DEATH EATERS RATE ON THE 14 DEFINING CHARACTERISTICS OF FASCISM?

I s Voldemort a fascist? In fact, is there a science to spotting fascists? It would be a very useful branch of science to master. And, once mastered, such a science could be weaponized to avoid fascists at all costs, death eaters or not. As far as empirical observation goes, the least subtle of the fascists are clearly easy to catalog: they might sport moustaches, wear Hugo Boss uniforms, or favor torch-lit rallies under the cover of darkness. But our taxonomy should also include the lesser-sported fascists, the ones that hide in plain sight. They might even seem to appeal to the darker side of your nature, until you begin to get a critical grip of your faculties. In short, not all fascists invade Poland, wear skulls on their caps, or talk incessantly about building walls.

So, what about Voldemort and his death eaters—were they fascists? They *seemed* like it. But how can we really tell? To help answer this question, it's useful to consider a fourteen-point definition of fascism that has been developed from a careful socio-political study of fascist regimes over the last century or so. The definition was developed by the scholar, Dr. Lawrence Britt, who examined the fascist regimes of Hitler in Germany, Mussolini in Italy, Franco in Spain, Suharto in Indonesia, and several Latin American regimes. From his study, Britt identified what he called the 14 defining characteristics common to each. So, let's set them out below and see how Voldemort and his death eaters compare.

I. Powerful and Continuing Nationalism

Fascist regimes have a habit of constantly using patriotic mottos, slogans, symbols, songs, and other such paraphernalia. Flags are ubiquitous. They are used everywhere as symbols on clothing and in public displays.

In *Harry Potter and the Deathly Hallows*, Voldemort and the death eaters overthrew the Ministry of Magic. After doing so, the death eaters built a huge statue in the main entrance of the Ministry emblazoned with the slogan, *Magic is Might*, as a testament to the power of pure wizard kind. The ongoing use of propaganda continues later in the book when Harry broke into Umbridge's office, and came across wizards printing pamphlets entitled *Mudbloods—and the Dangers They Pose to a Peaceful Pure-Blood Society*.

And who could forget the death eater symbol, the dark mark, that appears as a glittering green skull, "etched against the black sky like a new constellation," and with a snake protruding from its mouth? The dark mark was also worn on the inner part of the left forearm of the death eaters, appearing as a faint mark, similar to that of a vivid red tattoo when inactive, and jet black when active. British author, Christopher Hitchens, noted the significance of the lightning bolt in an article written in the *New York Times*. Hitchens pointed out that the bolt, which was the shape of the scar Harry received as a result of Voldemort's curse, and considered to be emblematic of the book series, was also the symbol of Sir Oswald Mosley's British Union of Fascists, a prominent group of Nazi sympathizers during the 1930s and 1940s. The Nazis themselves, in their SS, also made use of the symbol.

II. Disdain for the Recognition of Human Rights

Fascists cultivate an acute fear of enemies and an irrational need for security. Their regimes persuade people that human rights can be ignored in certain cases, out of alleged necessity. Populations are encouraged to "look the other way" from, or even approve of, torture, executions, assassinations, and long incarcerations of prisoners.

and

III. Identification of Enemies/ Scapegoats as a Unifying Cause

Fascist populations are often rallied into a unifying patriotic frenzy over the need to eliminate a perceived common threat, or foe. The alleged threats are typically of racial, ethnic, or religious minorities, or, if political, are usually dubbed liberal, communist, socialist, or, increasingly, terrorist.

In the Harry Potter Universe, we have "The Muggle-Born Registration Commission." When they snuck into the Ministry, Harry, Ron, and Hermione encountered evidence of this Commission. Voldemort's Ministry of Magic saw only pure-blood wizards as worthy of access to magic and a place in the wizarding community. We also discovered that muggle-born or half-blood wizards were being interrogated and routinely imprisoned in Azkaban. There was also the interrogation of Mary Cattermole, under the presence of the dementors, who was being questioned over the identity of the wizard from whom she stole her wand. Elsewhere we witness groups of muggle-born wizards on the run from the Ministry, and the abduction and murder of Hogwarts's muggle studies teacher, Charity Burbage, in front of a den of death eaters.

J.K. Rowling stated that the terms pure-blood, half-blood, and muggle-born compared to, "some of the real charts the Nazis used to show what constituted Aryan or Jewish blood. I saw one in the Holocaust Museum in Washington when I had already devised the pure-blood, half-blood, and muggle-born definitions, and was chilled to notice the similarity."

IV. Supremacy of the Military

Even when there are pervasive economic problems, the military is given a disproportionate amount of government funding, and the domestic agenda neglected. Soldiers and military service are glamorized.

Clearly, there's no military wing of wizard kind. And yet, when we learn about the Muggle-Born Registration Commission, we note the unusual presence of dementors at the associated trials. Previously, dementors were strictly controlled and limited to the dark towers of Azkaban. But, as the fascistic power of Voldemort increased, the Ministry slowly lost control of the Dementors, who turned up to torment and terrorize Harry and Dudley in Little Whinging. By the time of *The Deathly Hallows*, the Dementors were 100 percent aligned with Voldemort, and were present in the Ministry itself. They were a constant symbol of hopelessness and oppression, and their presence kept the terrified wizarding community in tow. They were more than just a threatening magical force, they were the shock troops of dark magic, and their presence and nature helped suck the joy and hope out of wizard kind.

A secondary quasi-military magical force was the Snatchers. They rounded up muggle-borns on the run, and handed them over to the death eaters for rewards. The Snatchers captured Harry, Ron, and Hermione and took them to Malfoy manor.

V. Rampant Sexism

Fascist governments tend to be almost exclusively male-dominated. Under such regimes, traditional gender roles are rendered more rigid. Divorce, abortion, and homosexuality are suppressed, and the state is represented as the supreme guardian of the family institution.

There is no overt sexism in the Harry Potter Universe. However, there is a strong implication that Voldemort and the death eaters are guardians of the family institution, as long as your veins carried magical ability and the right kind of blood. They seemed obsessed with the protection of traditional pure-blood families and eliminating muggle-born wizards who "sully bloodlines." In the questioning of Mrs. Cattermole, Dolores Umbridge showed no regard for Reg and Mary Cattermole's children, as their mother was muggle-born. But, the Weasleys were still allowed to work and live within the wizarding community (even though they were known to be members of the Order of Phoenix and allies of Dumbledore) because they were a pure-blood family, and were protected, as they could have been converted to the cause.

VI. Controlled Mass Media

Very often, fascist governments directly control the media. But the media can also operate under a veneer of liberty, while being indirectly controlled by government regulation, or sympathetic media spokesmen of corporate monopolies. (As American author Upton Sinclair once said, "Fascism is capitalism plus murder.") Consequently, censorship, especially in time of war, is very common.

The only real wizarding media in the Harry Potter Universe was the *Daily Prophet*. As the story developed from book to book, we saw the power of Voldemort grow, and the infiltration of the Ministry evolve. The political evolution of this narrative is reflected by the way in which Harry and Dumbledore are portrayed in the *Prophet*. After *The Goblet of Fire*, Cornelius Fudge pressured the *Prophet* to cover up Harry and Dumbledore's claim that Voldemort had returned. By the end of *The Order of the Phoenix*, however, when Voldemort revealed himself to the world at the Ministry, the *Prophet* finally broke the story. And yet, after the fall of the Ministry at the start of *The Deathly Hallows*, Dumbledore was once more portrayed as a crazy old fool, and Harry became "Undesirable No.1."

VII. Obsession with National Security

Fear is used as a motivational tool by the government over the masses. The Harry Potter Universe has an interesting take on the NSA. In *The Deathly Hallows*, the Ministry placed a kind of magical trace, known as a taboo, on the name Voldemort, so that if anyone spoke the name, the Ministry was immediately notified of the transgression. The fascistic power logic behind the taboo was the hope that the majority of wizard kind would be too scared to speak Voldemort's name and instead defer to the obedient and pusillanimous use of the phrase, "he who must not be named." Indeed, as Dumbledore said to Harry early on in the series, "Fear of the name increases fear of the thing itself." Later, and in defiance of power, Order of the Phoenix members started speaking Voldemort's actual name.

VIII. Religion and Government are Intertwined

Fascist governments tend to use the most common national religion as a tool to manipulate public opinion. Religious rhetoric is commonly spouted by government spokesmen, even when the major tenets of the religion are diametrically opposed to the government's policies or actions.

There is no wizard religion, as such. And yet the rhetoric spouted by Voldemort and the death eaters was steeped in a conservative obsession with the history and legends of the wizarding world. They governed with a ferocity of fear and intimidation that echoed the attitude and atrocities of the Spanish Inquisition.

Voldemort's choice of Horcrux objects was also heavily reliant on relics that felt religious in their significance. His fixation with the elder wand, one of the deathly hallows and an ancient legend in the wizarding world, had a kind of faithful significance. While these relics were not used to govern, these ancient legends were the closest thing that the wizarding world had to a religion.

IX. Corporate Power is Protected

Echoing Upton Sinclair's contention, Lawrence Britt found that the industrial and business aristocracy of a fascist nation are very often those who put the government leaders into power, creating a mutually beneficial business/ government relationship and power elite.

We learned very little about corporate power in the wizarding world. And yet the very pure-blood ideology of the death eaters was allied closely with old money and aristocratic wizard families. Both the Blacks and the Malfoys were examples of exceedingly wealthy and aristocratic wizard ancestry. We could easily attribute the pure-blood families' choice of fascistic death eater ideology as a way to protect family money, as well as pure wizard bloodlines.

Indeed, this attitude of aristocratic moneyed power contrasts sharply with the outlook of Harry and the Weasleys. Lucius Malfoy's moneyed influence looms over both the Hogwarts board of governors and the Ministry of Magic. The Malfoys constantly draw attention to their wealth in comparison to the Weasleys, whom they look down upon as "blood-traitors," whereas Harry's family is incredibly wealthy, but also generous and kind. And Sirius Black is thrown out of his family home for refusing to conform to the death eater ideology, with which the rest of the Black family agrees.

Also on the question of corporate power and money, in that *New York Times* article by Christopher Hitchens, he notes that, "The prejudice against bank-monopoly goblins is modeled more or less on anti-Semitism, and the foul treatment of elves is meant to put us in mind of slavery".

X. Labor Power is Suppressed

Because the organizing power of labor is the only real threat to a fascist government, labor unions are either eliminated entirely, or are severely suppressed.

Reference to labor power in the books is rather oblique, and yet interesting. Clearly, there was a wizard working force at the Ministry, at the small businesses of diagon alley and hogsmede, at the *Daily Prophet* and the Quibbler, and at St. Mungo's Hospital. But the attitude of the death eaters to honest labor was truly revealed when Voldemort came to power and we saw the kind of malice that the likes of Voldemort and Umbridge held toward magical creatures, such as the centaur, Firenze, and the disdain with which most aristocratic families treated goblins and house elves. To some extent, this attitude to magical creatures extended to other wizard families.

When Barty Crouch mistreated Winky in *The Goblet of Fire*, Hermione objected that such cruelty was akin to slavery. And yet Ron merely maintained that house elves like to be treated that way, and that they liked their hard work. Hermione was the only character that wanted to address such prejudice, until Voldemort's regime began to treat muggle-borns in the way that the rest of the community had always treated magical creatures.

XI. Disdain for Intellectuals and the Arts

Fascist nations tend to encourage an open hostility to higher education and academia. Typically, professors and other academics are censored, or even arrested. Similarly, free expression in the arts is openly attacked and progressive art banned. For example, the Nazis used the term 'degenerate art' to describe modern art, on the grounds that it was un-German, Jewish, or Communist. Instead, they promoted art that exalted racial purity and obedience.

The best example of such disdain was Umbridge's stint as High Inquisitor at Hogwarts. In that position, she did everything she could to limit the free-thinking education of the pupils, and censored or punished any students or professors who stepped out of line. Although it was never crystal clear whether Umbridge was a death eater, she eventually became one of the most avid supporters of Voldemort's ideology in the Ministry and is widely regarded as one of the most fascistic and despised characters in the series.

Scorn was poured upon divination and care of magical creatures courses. Hogwarts doesn't teach any arts as such, and so these fringe topics, and their eccentric and unorthodox professors, came under attack from Umbridge, leading to the firing of Hagrid and Trelawney. Umbridge also monitored and censored other classes at Hogwarts, and banned all gatherings and extra-curricular activities. McGonagall and transfiguration remained relatively unscathed, though this had more to do with McGonagall, who puts Umbridge in her place as much as possible, and is never intimidated by her.

XII. Obsession with Crime and Punishment

Under fascist regimes, the police are given almost limitless power to enforce laws. The people are expected and encouraged to condone police abuses and even forego civil liberties in the name of patriotism. National police forces with virtually unlimited power are common in fascist nations.

Punishment was close to the heart of fascist extraordinaire, Dolores Umbridge. She forced Harry to etch a corrective message upon his own flesh with a quill that scarred his hand. When Harry broke into Umbridge's office, he found a picture of himself with a note attached that obsessively declared, "to be punished."

Umbridge also headed up the Muggle-Born Registration Commission, and in *Order of the Phoenix*, set up an Inquisitorial Squad at Hogwarts. The play out of events within the confines of Hogwarts acted as a prescient microcosm of what happened later in the Ministry in *Deathly Hallows*. By this late in the narrative, the dementors, normally limited to Azkaban and are proven criminals themselves, are let loose to roam among the wizard population in the name of crime and punishment, with muggle-borns being sent to Azkaban for "stealing magic."

XIII. Rampant Cronyism and Corruption

Fascist regimes are typically governed by groups of friends and associates who appoint each other to government positions. They then use governmental power and authority to protect their friends from accountability. Typically, national resources and even treasures are appropriated by the regime, or even stolen outright by government leaders.

and

XIV. Fraudulent Elections

Elections in fascist nations are very often a complete sham. At other times, elections are manipulated by smear campaigns or even assassination of opposing candidates. Other tactics are the use of legislation to control voting numbers or political district boundaries, and manipulation of the media. Fascist nations also typically use their judiciaries to manipulate or control elections.

These last two characteristics are connected, especially in a small wizarding community where everyone seems to know one another. Somewhat like members of the British aristocracy in real muggle life, the death eaters form an incestuous and nepotistic network, as most know one another from school and family. For example, the Black and Malfoy wizard clans are both pure-blood families from Slytherin house and are closely related.

The first Minister for Magic was appointed in 1707 when Ulick Gamp was elected. Subsequently, each serving minister was democratically elected through a public wizard vote, and there was no fixed limit to a Minister's term in office. However, regular elections were held at a maximum interval of seven years. To subvert this democratic process, Voldemort took over the Ministry by placing wizards in key positions under the *Imperius* curse.

Voldemort did this to Pius Thicknesse, the Minister of Magic at the time the death eaters infiltrated the Ministry. Thicknesse was placed under the *Imperius* Curse, and appointed Minister following the coup. Effectively a puppet of the death eater regime, Thicknesse was unconscious of his actions, and was consequently omitted from most official records as a Minister.

PART II

TECHNICAL TRICKERY AND PARAPHERNALIA

COULD SCIENTISTS BE THE MODERN WIZARDS?

"Any sufficiently advanced technology is indistinguishable from magic."

This quote comes from Arthur C. Clarke, the British futurist writer, maybe most famous for co-writing the screenplay of the 1968 movie, *2001: A Space Odyssey*, widely thought to be one of the most influential films of all time. The quote is one of Clarke's famous Three Laws, the other two of which are also relevant to thoughts about magic: "When a distinguished but elderly scientist states that something is possible, he is almost certainly right. When he states that something is impossible, he is very probably wrong," and, "The only way of discovering the limits of the possible is to venture a little way past them into the impossible." So, does Clarke's famous Third Law make scientists modern wizards? To answer this question, it'll help to consider the relationship between magic and science.

What is science? Like magic, science is a recipe for doing certain things. And, like magic, science is ancient. It developed over many thousands of years, and passed through many cultures and societies, evolved through many metamorphoses. In classical times, science was merely one aspect of the work of the sophist. In medieval times, science was an elemental feature of the work of the alchemist, or the astrologer.

To help compare science and magic, we can think about the four pillars upon which both are based. Like magic, science is a worldview; science is an institution; science is a method; and science is a body of knowledge. To consider whether scientists might be the latter-day wizards, let's focus on science as a worldview and a method.

Science as a Worldview

Magic and science share a common origin. Science's worldview is one of the most powerful influences that have shaped our attitudes to the universe. It can be traced back to early forms that held some considerable sway in antiquity. Science's tradition, which links it strongly to technique, was the knowledge passed from craftsman to apprentice, elder to novice, exists from the earliest of societies and cultures. This tradition began long before science developed as a method, discrete from everyday practice and folklore.

In early times, humans sought to control nature. The primeval world was profuse with a huge number of plants and animals, which varied widely as humans went on migratory journeys. We were parasitic on uncontrolled nature. So humans needed techniques we could use to try understanding nature, as any mistakes could often be fatal. The preservation and propagation of fire, for example, led to the very simple and essentially chemical technique of cooking. The observation of plants and the habits of animals laid the basis of biology, and the spoils of the tribal hunt would have led to a rudimentary knowledge of anatomy. But by hunting and gathering and observing, technique could only go so far.

Magic evolved to fill the gaps left by any early limitations in technique. Humans used animals as magic totems. The tribe would use images of the totem, or maybe symbols and even dances, to encourage the animal to prosper and multiply. A human animagus, we might call them early magicians, would essentially become the animal. As long as the rules of the totem were followed, the tribe would flourish.

The totems became associated with certain powers. Perhaps they were sacred, or taboo. They had to be handled carefully, or else the balance of nature would be upset. The totem carried a certain *mana*, or power, which signifies its influence over humans. Such totemic symbols exist to this day, in the lion of Gryffindor, the serpent of Slytherin, the badger of Hufflepuff, and the eagle of Ravenclaw.

Theory of Magic Spirits

The methods of the early magicians were based on mimicking and sympathizing with the workings of the universe. Based on archeological evidence of the cave art of Western Europe, it seems these magicians were already established in the Old Stone Age. Take the cave paintings of the Trois-Frères in the Ariège department of southwestern France, for example. A painting there shows a magician or sorcerer wearing stag's horns, an owl mask, wolf ears, the forelegs of a bear, and the tail of a horse. The value of such an animagi could have been to ensure a successful hunt.

The magician or sorcerer, the cave paintings of the Trois-Frères in the Ariège department of southwestern France.

At first, the magicians would use likenesses, and later symbols, to perform an operation on something that would be considered transferable to the real world. An unbroken thread links these ancient symbols to those used with such success in modern science.

Another feature of primitive thought, which at some point separated itself from imitative or symbolic magic, was the idea of the influence wrought upon the real world by spirits. The idea of a spirit probably emerged from the reluctance to accept the fact of death. Early spirits were very worldly, members of the tribe who had since passed on. But the idea evolved that it was necessary to win, or regain, the favor of a spirit, now god, by doing something that pleased them.

The old idea of spirits split into two very different forms. On the one hand, it transformed into the idea of spirit as an all-powerful being, or god, that was to become the central figure in religion. And on the other hand, the spirit became divorced from human origin to become an invisible natural agent, such as the wind, or the assumed active force behind chemical and other crucial changes. This second idea of the spirit was to become hugely important in the evolution of the understanding of spirits and gases in science.

Witchcraft to the Ignorant

Science and magic are far more entangled than you think. At first the rituals of magic would have involved most of the tribe. But, in time, cave art shows solitary figures of the tribal animagus, dressed as an animal, who appears to have some special place. In many primitive tribes today, there are still medicine men, or magicians. They are held in high esteem, as they are thought to have a peculiar relationship with the forces of nature and the universe. To some extent, they are set apart from the normal work of the tribe. And, in return, they exercise their magical arts for the tribal good. They are keepers of learning and knowledge. They are the forerunner, the lineal cultural ancestor, of philosophers and scientists.

Arthur C. Clarke's Third Law echoed a statement from a 1942 story by Leigh Brackett, "Witchcraft to the ignorant . . . simple science to the learned." The accelerating pace of change seen throughout the 20th century and early 21st century overwhelms many. Those without science are disconnected from the science and technology of the age. To them, inexplicable science is the modern counterpart of magic, and the scientist the witch.

CAN SCIENTISTS EVER DEMONSTRATE WINGARDIUM LEVIOSA?

There is that jaw dropping moment in every muggle's life when they see something levitated without any obvious explanation of how, except for the possibility that the person causing it has some special, magical ability.

Muggle magicians and spiritualists have been amazing people with these impossible feats for centuries. Increasingly, though, the people who have been providing the most public awe have been scientists, whose exploits are equally as mesmerizing. These talented men and women perform their feats by enhancing their knowledge of nature and developing techniques to exploit it.

In Harry Potter, levitation is considered one of the wizard's most rudimentary skills. It can be achieved a number of ways, like with Wingardium Leviosa or the Locomotor spells, which can lift a target a few inches off the ground and move it in any given direction. In the real world, though, what methods have muggles mustered to achieve levitation?

The Gravity of the Situation

Levitation is all about somehow supporting the weight of an object in midair. When we typically talk about weight, we're used to saying that something weighs, say, 100 pounds, but when we say this, we are really talking about the mass of the object. The mass of an object doesn't change from place to place and is a measure of how much matter is contained within it.

All objects have a gravitational field associated with them that attracts other objects that have mass or energy. The bigger an object's mass, the larger the associated gravitational field and the more it attracts other objects. As the Earth has a very large mass compared to objects on it, it's gravitational field is the one that dominates objects on or near Earth. The force of attraction within Earth's gravitational field is proportional to the mass of the object being attracted to the Earth. This attractive force is what scientists refer to as the weight of the object.

To levitate an object on Earth, it's necessary to find a mechanism that can overcome the object's gravitational pull toward the Earth i.e. its weight. Either that, or somehow find a way to negate the effect of Earth's gravity, like H. G. Wells's fictional substance called cavorite from his 1901 novel, *The First Men in the Moon*. Unfortunately, in our non-fictional magic-absent world, such a substance or quality is deemed impossible, although it didn't stop some engineers and scientists from giving it a good go.

From the mid-1990s there was a BAE Systems supported research program called Project Greenglow, which was set up with the sole purpose of developing antigravity technology. The man behind the project, Ronald Evans, was inspired by the thought of gravity control and gravitational propulsion. Despite all the enthusiastic work throughout the following two decades, the project officially closed in 2005, with no real viable antigravity technologies on the horizon. Antigravity attempts aside, there are a few existing technologies that can overcome an object's gravitational pull toward Earth i.e. its weight. A relatively obvious one is through the use of air.

Aerodynamic levitation

If we needed to get a feather to stay aloft, we could simply blow it with puffs of air from underneath. Generally, the bigger and heavier the feather, the larger the puff of air needs to be to support it. However, its shape and orientation are important, too. This technique isn't very impressive, though.

A more impressive trick is to suspend a ping pong ball in a current of air from a bent straw or hair dryer. The ping pong ball sits on a cushion of air pressure provided by the upward air current, though the ball doesn't fall off of this stream of air like a feather eventually would. As the airflow moves around the ping pong, it creates regions of high and low pressure, which provides a restoring force that keeps the ball in the stream of air and balanced against its weight. It will even work with the airflow tilted a little to the side. If tilted too much, though, the weight overcomes the pressure and the ball drops.

By using a larger stream of air, it's possible to suspend a beach ball, but the application doesn't stop at completely spherical objects. Using more powerful streams of air, it's possible to suspend a screwdriver, egg, light bulb, test tube, and a small bottle, among many other things. It's even possible to keep a human aloft with air, although it involves a slightly different process.

When skydivers jump out of planes, they are falling past the air. As they fall faster, the air resistance (drag) on them increases until it balances the weight forces pulling them down. With the upward and downward forces in balance, the skydivers don't gain any extra speed and are said to have reached their terminal velocity.

If enough air is pushed upward at a speed equal to a skydiver's terminal velocity, then the downward force of their weight will be balanced by the upward drag caused by the moving air. This can cause the skydiver to be suspended in midair, which is how indoor skydiving centers work.

Although large and small objects can be levitated using aerodynamic forces, it probably wouldn't be very practical to blast objects with strong gusts of wind to keep them airborne. There would likely be too much noise and/or collateral damage. So, maybe this next levitation technique is better?

Acoustic Levitation

Acoustic levitation is a process that makes it possible to lift and move objects using sound waves alone. A sound wave consists of areas of alternating high and low pressure, where molecules have been pushed together (compression) or drawn apart (rarefaction) respectively. The distance between successive compressions or between successive rarefactions is called the wavelength, and however many pass a point every second indicates the frequency of the sound wave.

Sound waves can exert a pressure on surfaces they come into contact with. It's called the acoustic radiation pressure or acoustic radiation force. If the amplitude (volume) of the ultrasonic sound wave is large enough, it can carry sufficient energy to suspend things.

Acoustic levitation uses a loudspeaker or transducer to produce sound waves with frequencies higher than 20,000 hertz, known as ultrasound, because it's beyond the limit of human hearing. Like any sound waves, if these sound waves encounter other sound waves, they can interact and produce a resulting wave pattern that is a combination of the waves. If identical sound waves are traveling in opposite directions when they meet, this interaction or interference can result in a standing wave.

In standing waves, there are points which vary between maximum and minimum pressures, called antinodes. Then there are points that lie between the antinodes in which the pressure doesn't vary at all, called nodes. The nodes represent stable areas within the standing waves, in which objects can be levitated, providing they are small and light enough. Objects are usually limited to a size of a quarter to a half of the wavelength of the sound waves, which can be about 17mm or less. So, objects larger than about 4mm generally can't be supported in this type of ultrasound standing wave.

Using this system, polystyrene balls and droplets of water have been levitated, as well as ants, ladybugs, and tiny fish. Other systems use multiple loudspeaker sound sources to manipulate objects like small screws, matchsticks, and LEDs. By controlling the sound output from each individual loudspeaker, the objects can be levitated in stable regions as well as moved around; similar to the locomotor spell.

So, what about levitating a feather like Hermione did in *Harry Potter and the Sorcerer's Stone*? Well, we put the question to Asier Marzo, who researches acoustic levitation at University of Bristol, UK. He'd never tried it before but within weeks of our request, he sent a link to a video in which, yes, a feather is successfully levitated. The feather was only about a centimeter or so across, but it proved that acoustic levitation could be used to produce effects similar to those seen with the levitation charm.

Diamagnetic Levitation

You may be familiar with the effect that magnetism has on materials that contain iron. Iron is known as a ferromagnetic material and is strongly attracted to magnetic fields. Cobalt and nickel are also ferromagnetic.

Ferromagnetism is the most familiar form of magnetism, but there are also others, such as paramagnetism and diamagnetism. Paramagnetic materials are weakly attracted to an external magnetic field, whereas materials with diamagnetic properties tend to be repelled by an external magnetic field.

Diamagnetism acts on all materials (rather than just metals), causing them to feel a relatively weak repulsion when in a strong enough magnetic field. This can cause objects to levitate when they are within a strong enough vertical magnetic field, as demonstrated with small frogs, crickets and mice.

When levitating a frog, it's not just a case of applying the strongest magnetic fields. Although this would provide lift to the frog, it wouldn't allow the frog to remain hovering, as instabilities can quickly set in. This is because there is a stable zone on the vertical axis within which diamagnetic objects can be levitated. To achieve levitation, the magnetic fields need to be adjusted to an accuracy of a few percent. The theory underpinning the flying frogs was developed in the 1990s by British professor Michael Berry, a leading theorist in mathematical quantum physics.

It is still a problem to create a stable zone big enough to accommodate a human being, though. You would need a magnet that uses about 100 megawatts of power and it would need to have a central space of 2 feet in diameter to levitate a human. For comparison, the experimental levitating space (at the center of the superconducting magnet) used to levitate the mouse had a diameter of 60 mm.

So, Can Scientists Ever Demonstrate Wingardium Leviosa?

The answer is most definitely yes. On a smaller scale, acoustic levitation can be used to levitate and manipulate objects, but it requires an array of transducers to send the waves and is limited to roughly 4mm objects. Diamagnetic levitation has improved on that by levitating bigger objects, including small animals; but only within particular stable regions and not yet for larger objects. It also requires a huge and powerful magnet to work. For larger objects, aerodynamic levitation is the most capable option, but it relies on a strong blast of air, so rather inconvenient again. In any case, levitation is a real thing and a very active area of research.

HOW HAZARDOUS IS A FLYING BROOMSTICK?

When it comes to personalized transport, muggles have it made. We've got skateboards, roller skates, segways, bicycles, motorcycles, cars, microlights, and even jet packs. Despite all this variety, we still haven't opted for a mode of transport that resembles anything close to a broomstick. Not surprising really, considering they look like the equivalent of riding a bike with the handlebars, seat post, and seat missing.

At first glance, you might think that riding a broomstick is one of the most uncomfortable aspects of a wizard's regimen. Not so. To avoid any posterior problems, they make use of a cushioning charm, which essentially provides a magical replacement for a seat.

Although, this solution may not have gone down well with the muggle versions of witches, who allegedly rode their brooms as a way to administer psychoactive potions via their private parts. Rather than flying while under the influence, these witches would have been flying because of the influence. Putting aside thoughts of saddle sores and high flyers, what hazards, if any, could broomsticks present as a mode of everyday transport?

Flight of Fancy

The notion of witches flying on brooms is more than 500 years old now. They weren't always depicted as flying with the bristles behind them either. Some witches flew with the bristles in front of them, similar to how a hobby horse is used. Just to clear things up from the start, an ordinary broomstick can't be used to shuttle a person through the air. Nothing about its shape, materials, or design give any impression that a broom would make a good flying device.

Traditional brooms, called besoms, consisted of a sturdy wooden pole with a bunch of twigs attached for sweeping. These twigs were often obtained from a shrub called broom (Cytisus scoparius), which is where our use of the word broom likely came from. Back then the broom twigs were mostly arranged in a cone shape, but from around 1800 broomcorn (Sorghum bicolor) bristles became popular, especially in America. These also tended to be formed into a flattened cone, particularly with a machine called a Shaker broom vise.

In Potter, the broomstick tails are made from the twigs of trees such as birch or hazel. On J.K. Rowling's *Pottermore* website it says that the "birch is reputed to give more 'oomph' in high ascents, whereas hazel is preferred by those who prefer hair-trigger steering". One seems to provide more power while the other provides more responsive turning. Using muggle technology, the only long and thin flying machines capable of these attributes are missiles and rockets.

A rocket is powered by the combustion of fuel with oxygen in its engine. The hot gases that are created are exhausted from the tail of the rocket. As the gases are ejected rearwards, they push back on the rocket in what's called an equal and opposite reaction. This propels the rocket forward. A rocket powered broom wouldn't make the best vehicle though. It wouldn't be able to fly horizontally with the weight of a person on it, unless its lift is generated some other way. Also, it would travel at break neck speeds while spewing raging hot flames from behind. Definitely a safety hazard.

One top of the range broomstick, the Firebolt, is described as, "The state-of-the-art racing broom. The Firebolt has unsurpassable balance and pinpoint precision. Aerodynamic perfection." It's also said to go from 0 to 150 mph in 10 seconds, which is about the same acceleration as a BMW S1000RR motorcycle.

At high speeds, motorcyclists tuck in to present as small a form to the oncoming air, and therefore reduce the aerodynamic resistance, or drag, on their body. This would likely be necessary on a broomstick too, where the rider appears to be fully exposed to any oncoming air. In fact, in comparison to the drag already present on the rider, the shape or aerodynamics of the broom wouldn't make the biggest difference, unless it has a charm that helps to shield the rider from any unwanted aerodynamic effects.

A nonmagical equivalent of that shield can be seen on motorcycles, where riders have an aerodynamic windshield that they can tuck behind as they approach significant speeds. For added safety they are also decked from top to toe in protective gear, something that only seems to be partly present on Quidditch players.

Keeping on Top of Things

Anyone who's ever tried to ride a bike will know that the first thing you have to learn is how to stay on the thing in the first place. It's not enough to just sit on it; a bike can only stay upright if it's in motion or if it has someone on it who is particularly skilled at keeping balanced. However, the ground offers some support so the rider just has to put their foot down to stop the bike from tipping to the left or right, as bicycles lack lateral stability. What about a broomstick?

The main difference is that a broomstick appears to be suspended in midair where it is free to rotate in all directions. It can rotate left and right (called yaw), up and down (called pitch), or roll to the left and right. Without anything to support it, the broomstick would be rather unstable, requiring a great deal of balancing skill from the rider. But seeing as Neville Longbottom was able to maintain balance on his first broom ride (granted, it was a hazardous experience), there must be something inherent in the broomstick to aid in balancing, such as some particular charm or other magical effect. For example, the Firebolt broomstick is described as having "goblin-made ironwork (including footrests, stand, and twig bands) . . . which seem to give the Firebolt additional stability and power in adverse weather conditions."

Built-in stability isn't just a magical invention though. Muggles have built a variety of machines that have that ability, but instead of a magical charm, muggles have instead made use of computer programming. A prime example is the personal transportation device called the Segway, which has the ability to keep itself balanced with or without a rider. It does so with the help of complex programming embedded within its components. Pretty much scientific magic.

So, in some ways a broomstick can be likened to technologies present in bicycles and Segways. The broomstick is mounted in a similar way to a bicycle whilst having the ability to keep itself upright and balanced, like a Segway. However, despite any inbuilt stability, it's still necessary for riders to keep themselves balanced on the vehicle while in transit. Any humps, bumps, or sudden changes in speed or direction could throw a rider off their mount, especially at higher speeds. Of course, there's also the possibility that your vehicle may have been sabotaged or even "broom jinxed."

What's the Worst That Could Happen?

The consequences of losing control generally become more severe at higher speeds. If you're not directly injured from a high-speed collision with a person, wall, or rogue bludger, the resulting impact with the ground will surely give you something to think about. Although, as the saying goes, it's not the fall that'll kill you. It's the sudden stop.

For muggles, personal vehicles don't only pose a risk to the person piloting them. For example, in 2015, 70,000 US pedestrians were injured and 5,376 lost their lives as a result of motor vehicle crashes. And in Britain from 2011 to 2015, cyclists were involved in around 1 percent of pedestrian fatalities. It pays to remember that nearly all vehicles can become dangerous weapons when up against the vulnerability of the human body. Again, the faster or bigger the vehicle, the more potentially life threatening the consequences can be.

Surely a broomstick wouldn't cause a great deal of harm though. What specific dangers could they pose? Well, staying well clear of urban legends about deaths by broomstick, there was a case in 1888 in South Wales, UK. It involved an argument between two ladies in which a broomstick was thrown in temper, instantly killing the victim. The post mortem determined that she had suffered a fracture to the base of her skull. The offender was subsequently found guilty of 'throwing a broomstick with provocation'.

Then, in 1901, Los Angeles, a young girl was impaled by a broomstick. She was playing on a haystack with some friends and fell off, landing on the upturned broom and ending up with nearly a foot of it penetrating her abdomen. The report said the accident would probably cause her death, but regardless of the outcome, it just goes to show how dangerous a broomstick can be.

Seen in this way, a game of Quidditch could be fatal, although it's been stated by J.K. Rowling that people rarely die playing the game. There is a part in *Harry Potter and the Chamber of Secrets* where Harry Potter is chased by a "rogue bludger". In the movie version, we saw him swoop through the screaming crowd while they ducked out of the way. Judging by the damage a broomstick can cause when someone falls on it, the potential outcome of this chase could have been catastrophic.

Having said all this, wizards have powerful spells that can instantly heal many unfortunate injuries. In the muggle world, doctors are equally as amazing although the healing takes a lot longer. One man had a broomstick pierce through his cheek and down into his collarbone. Although scarred, he is currently alive, well, and cheerfully recounting his tale.

Safety Hazard?

So, assuming that muggles could get a broomstick to fly, we'd have to make sure the propulsion method wasn't a hazard in itself. For example, rockets are a no-go. And regardless of any potential internal stability it may have, the rider would need a good ability to balance on the stick, including a good grip to avoid falling to their death. Air resistance would also be a battle, so they would need to adopt a good aerodynamic posture to appear more streamlined when travelling at speed.

In case of losing control of the broomstick, there would need to be safety measures in place to protect anyone from getting impaled by it. This could be something like armor for anyone in the vicinity (like spectators at a Quidditch match). Even better, there could be a dead man's switch to automatically stop the broom in case a rider somehow loses control or falls from the broom. In any case, it appears that flying broomsticks would be very hazardous indeed.

Something to think about if you're ever offered a front row seat at a Quidditch match.

CAN MUGGLES MAKE A MOTORCAR THAT FLIES?

You're sitting with a friend by the Post Office Tower in London, having a relaxing day in town. It's a lovely day, the sky is blue, then a glint catches your eye. You think it's a low flying plane, but hang on. That's no plane. It looks more like an old car!

Shocked, you turn to your buddy, exclaiming, "Dude! There's a car flying through the air!"

But when they look up, it has mysteriously disappeared. They don't believe you at all and suggest you have a word with the Department of Intoxicating Substances. But you're certain of what you saw and are determined to delve deeper. So, in the absence of a magical explanation, you take out your phone to see what the World Wide Web has to say on the matter.

Cars That Fly

The first real automobiles were heavy, steam-powered vehicles. This changed in 1807 when the first internal combustion engines appeared. Today, these engines are still the most widely used power source for cars. In 1903, almost a century after the first combustion engine, the Wright brothers famously took to the air to successfully demonstrate powered flight.

Within fifteen years of the Wright brothers' achievement, a man named Glen Curtiss developed the Curtiss Autoplane. It was essentially the wings of a triplane attached to the body of a motorcar along with a tail and rear-mounted propeller. This would have been the world's first known flying car—if it actually flew, that is. Apparently, it only managed a few short hops before World War I took precedence.

Since then, there have been many successful vehicles of this type; referred to as roadable aircrafts. Some of these were modeled on planes that upon landing could detach the wings, propeller, and tail to become a road vehicle. Others were basically cars that could have flight components added to make them into aircrafts. If you saw one of these flying in the air, the only words you could justifiably use to describe it would be, "Look, it's a flying car!" If you have any doubt on this matter, check out the ConvAirCar which flew in the mid-1940s, or the AVE Mizar, whose inventor unfortunately died after crashing it in 1973.

There have been quite a few flight-related accidents on the road to roadable aircrafts. Twenty years earlier, Leyland Bryan, the designer of the Bryan Autoplane, died after one of his vehicle's wing sections hadn't secured properly for flight. This particular roadable aircraft was different in that its wings could be folded up around the plane for ground travel, saving space and eliminating the dependence on a hangar to store the flight components. Fortunately for Harry Potter and the Weasleys, the Ford Anglia did not have any wing sections that could fail, as it flew by other means, although Harry falling out of an unlocked door could have potentially been just as perilous.

On a Wing and a Prayer

Despite the possible dangers, muggle inventors haven't been put off. Nowadays, the most promising flying car candidates are pretty much just more sophisticated versions of the mid-20th century attempts. For example, the Terrafugia has wings that fold up on its sides, similar to the Bryan Autoplane, while the new Aeromobil uses wings that fold back along its length but tuck in like the wings of a wasp.

What the above vehicles have in common is their use of wings to achieve lift for flight. This was and still is the most common way to get a motor vehicle to fly. However, a problem with wings is that they need a certain length of runway for takeoff and landing and they also need a roadway wide enough to accommodate the wingspan.

For takeoff, the flying vehicle must attain a particular speed, known as the takeoff speed. It takes the vehicle a certain amount of takeoff distance to accelerate from a standstill up to the takeoff speed. At this takeoff speed, the air moving over the wings can provide a sufficient lifting force to overcome the weight of the vehicle, causing the craft to be lifted skyward.

However, once airborne, a winged roadable aircraft can no longer use its wheels to maintain its forward motion, so the vehicle must switch to an alternative source of forward thrust. This is why they nearly always feature propellers. If the vehicle slows down, its wings will produce less lift, but if it slows down too much, the lift produced will no longer be able to support the vehicle's weight. Simply put, the vehicle will return to the Earth either in a glide or a crash.

If the Weasley brothers had opted for a less magical approach and obtained a winged flying Ford Anglia, the rescue of Harry from the Dursleys would have gone down quite differently. Firstly, they certainly couldn't have hovered outside Harry's bedroom window like that. Although, I'm pretty sure there are laws against that sort of thing. Secondly, they would have had to land on the road at the front of the house, presuming the road was wide enough to accommodate the wings without damaging parked cars, street lights, or the neighbor's tremendous topiary.

So, although wings work, they still have their drawbacks. What we want is a flying car that can hover, and the best candidates for that are vertical takeoff and landing (VTOL) aircrafts.

VTOL

Helicopters are the most recognized VTOL aircrafts, but they aren't so good for traveling along roads or hovering outside a row of terraced houses, especially with their hazardous open blades. As such, more attention has been focused on vehicles with contained blades, either in the form of ducted fans or propellers.

Ducted fans are basically fans that are housed in a cylindrical section or duct. These can provide both lift and propulsion by altering their angles directly or else using flap-like blades to redirect the direction of the air exiting the fan. They can also be quieter, safer, and more efficient than non-ducted fans at lower speeds, while having a higher thrust-to-weight ratio. This is handy for aircraft design in which a key goal is the reduction of weight without limiting performance.

An example of an aircraft that uses this technology is the futuristic-looking Moller M400 Skycar. The M400 has only ever flown while tethered to a crane for safety reasons, and even then, it did not have an onboard pilot but was instead operated via remote control. A major issue was the stability of the aircraft, but the company is currently focusing on its other flying vehicles. It's designer, Paul Moller, hopes that in the future, vehicles like the M400 could be used in rescue situations. For example, they could pull up next to a building to allow a person in peril (or imprisoned in a bedroom) to climb on board and be ferried away to safety.

So, personal VTOL flying cars *do* exist—in prototype form, at least. But, to become commercial vehicles, they first have to obtain the necessary certification, which is obtained from the Federal Aviation Administration (FAA), whose stated mission is to provide the safest, most efficient aerospace system in the world.

Ron the Rebel

Across the globe, muggles have developed strict transport regulations that must be adhered to by vehicle manufacturers. In regard to flight, these regulations are moderated by national aviation authorities such as the Civil Aviation Authority (CAA) in the UK or the Federal Aviation Administration (FAA) in the US. The presence of these organizations helps to ensure consistent levels of safety and consumer protection.

A hope for flying cars is that they might one day become as common a form of transportation as cars or buses are today. These modes of transport are relatively easy to get licenses for, considering that in 2014 there were 45.5 million active driving records in Great Britain. However, to use a flying car you would need a pilot's license, which requires a larger investment of time and money.

While Ron thinks it's a good idea to effectively take his father's Ford Anglia 105E for a joyride, he's breaking muggle laws by driving under age and without the proper license. Not to mention putting his and Harry's lives in jeopardy before (ahem) emergency parking in the Whomping Willow. But a license doesn't guarantee total safety, anyway. Even though licenses are needed to fly and drive, accidents are still inevitable whether due to human error, technological error, or an act of God (if not an act of wizardry).

To overcome the potential for human error and the need for flying car owners to have a private pilot certificate, autonomous vehicles are a desirable option. Google, Uber, and Tesla are some of the companies pushing this technology, which once established would no doubt become a necessary feature of any future flying car network. This would also leave space for a flying car to travel around without an on-board pilot, similar to the 105E after it ejects Harry and Ron, or when it rescues them. So, here, technology is reproducing something that would normally only make sense in a magical world.

Conclusion

Quick study finished. You can put your phone away and assess the options from what you have just seen. There were no visible wings or propellers, and if it used ducted fans, it must have been operating an unheard of silent type. At this point, the tentative conclusion of an unidentified flying object (UFO) would have worked, except your online search for similar looking cars has already identified it as a possible 1960's Ford Anglia 105E.

Remembering recent reports and rumors of strange goings-on locally, you conclude that there must be something happening that is as yet unknown to mainstream muggle science, or maybe your friend was right about the intoxicating substances and you need to ease off the butterbeer.

In either case, flying cars really do exist, but there are many obstacles to achieve before flying cars can launch from our front yards. The main problems include finding one that complies to legislation and is considered both road and airworthy, as well as having a nationwide infrastructure to support the needs of flying motorcars such as takeoff, parking, and air traffic control. There is also the possibility that you were looking in the wrong place. Maybe the forbidden forest would have been a better place to look.

COULD SCIENCE DEVELOP MOODY'S MAD EYE?

Who gets your vote for the most badass wizard? Other than Dumbledore, Voldemort, and Snape, naturally. Is it Sirius Black, perhaps? Sirius, the animagus, looked like a cross between Ozzy Osbourne and Francis Ford Coppola's *Dracula* (also played by Gary Oldman). Or maybe it's Gellert Grindelwald, the brilliantly talented (and brilliantly-named!) wielder of the Elder Wand, and known as the Second Most Powerful Dark Wizard of All Time. That's some title. Or perhaps your pick of the wizardry crop is Alastor Mad-Eye Moody.

Considered by many to have been the most powerful Auror of all time, Mad-Eye was a master of magic, both offensive and defensive. In the first wizarding war and its aftermath, Mad-Eye battled and defeated dozens of deadly death eaters. And Voldemort considered Mad-Eye to be such a deadly enemy that he made him his prime target, of all the talented wizards and witches shielding Harry during the Battle of the Seven Potters.

Now Moody was most famous, of course, for his magical eye. A magical prosthetic that replaced an eye lost in battle, the eye was an electric-blue orb that sat in his empty eye socket. The eye was able to rotate through a full 360 degrees in Moody's head, and enabled him to spy through anything, whether it was wood, cloaks of invisibility, or even through the back of his own mad bonce. Indeed, the eye seems to have been made just for Moody, as when Barty Crouch Jr. wore it, the eye got stuck in mid-whirl.

But what was the eye's full range of function? In the movie version of *The Goblet of Fire*, the eye was seen to have a zoom function. And the fact that the eye was powerful enough to see through the Cloak of Invisibility, one of the Deathly Hallows, suggests that it may have been a very rare artifact indeed. The Cloak, according to legend, granted the owner true invisibility. As the origin of the eye was never known, it's possible the eye was itself an ancient and very powerful artefact, matching the Deathly Hallows if not in fame or mythical status then certainly in power.

The Muggle Mad-Eye

So, what are the chances of a muggle Mad-Eye? We already live in an age where muggle technology can help restore vision to the blind. The science is simple enough. The tech works by a combination of an external eye-glass-mounted camera with a complex retinal implant. The micro-chipped camera interprets what it sees, and wirelessly runs the visual data to the implant, which houses 60 electrodes to feed information to the optic nerve, the nerve that discerns light, shape, and movement. The resulting vision isn't quite the same as typical sight. The muggle sees contrast, and the edges of things, but in black and white only. The damaged cells in the eyes disable the natural ability to see light and color. But, with use, the brain can learn how to make sense of the signals, and convert them into images. Users are then able to read books, cross the street, or see images of their kids for the first time in years. Not only that, but scholars the world over are now working to enhance what they call a retinal prosthesis system, or a muggle Mad-Eye. The next generation of eyes will be able to see color, by using improved algorithms that gauge electrode data, and the device will give sharper images, enabling the kind of eyesight focus you get on computer screens, which can alter resolution and brightness. New muggle eye models have more electrodes, which means better resolution, and many eye researchers predict a fully functioning synthetic eye on the market by 2020.

The next big thing in eye tech will be to bypass the eye and go straight to the brain. This revolution in technology, where implants bypass the retinal layer and go directly into the visual region of the brain could mean a huge breakthrough for millions of visually-impaired muggles. The new eye tech may not be able to see through wood, cloaks of invisibility, or the back of a muggle head, but the device may enable a kind of superhuman ability, including telescopic sight, just like the zoom on Moody's eye.

Muggle brains can be taught how to interpret the powerful zoom functions of cameras. And that means muggle eye wearers could learn to see much closer or farther than the normal human eye. But it gets even more bionic. The muggle eye might be able to see more of the electromagnetic spectrum. That would not only include the visible region we are used to (from red, through orange, yellow, green, blue, and indigo, to violet), but also infrared. And that would mean an eye capable of heat-sensing ability, the capacity to detect some gases, and yes, even the ability to see through objects!

So, in the future, muggles might become a little like cyborgs, or walking science labs. Through a prosthetic eye, we might have available a whole range of apps and devices. An x-ray vision app might enable muggle military recruits to detect landmines on the terrain of battle. Doting parents might have an eye app that enables them to detect toxic gases in their kids bedrooms, in a similar way to the way carbon monoxide alarms work.

The muggle eye may even go beyond the magic of Moody's eye. As images would be projected directly into the visual areas of the muggle brain, we may see things we never imagined. We might envision the millions of creeping microbes living on the human body. As the eye will never sleep, it could be set to guard us at all times of day and wake you at night if danger looms or light dawns outside. As the muggle eye would be Wi-Fi enabled, the wearer could record their day-to-day life and beam it straight online. Uh, oh. And your favorite movie or TV show could be streamed straight to your brain.

Currently, the muggle eye sees around 1 percent of the electromagnetic spectrum. That's not a lot of this big old universe, when you think about it. But, in a future where muggle bodies are augmented by a device such as the eye, our experience of the cosmos will be utterly transformed.

WHEN WILL MUGGLES DEVELOP MOVING PORTRAITS?

"We have discovered nothing," Pablo Picasso once said of modern art. The great Spanish painter, and the co-inventor of collage, was speaking in 1940 as he emerged from the newly-discovered Lascaux caves in the Dordogne. Carbon isotope analysis of the charcoal used in pictures of horses at Chauvet, south-central France, had shown that prehistoric cave art was at least 30,000 years old, a discovery that prompted Picasso's famous rethink about the progress of art. Because the horse paintings were just as artistic and complex as the later Lascaux paintings, it had indicated that art developed much earlier than had been realized. The cave art found in parts of France and Spain had shown ancient man to be a remarkably talented artist.

But perhaps Picasso would change his mind if he'd seen magical portraits. Magical portraits could walk and talk. Their subjects might even have moved from frame to frame. Good muggle portrait art depends on the charisma of the sitter and the skill of the painter to bring them to life. But magical portraits took it to another dimension—they actually moved and behaved like their subjects. And the extent to which the magical portrait interacted with the viewer very much depended not on the talent of the artist, but on the power of the magical subject depicted.

Muggle art is meant to capture the essence of the sitter—magical art went one step further. When a magical portrait was taken, some of the sitter's essence, perhaps their favorite phrases and definitive deportment, was captured to ensure the painting was a true representation. Witness the portrait of Sir Cadogan, who was always challenging the viewer to a fight, or was forever falling off his horse. Or the Fat Lady portrait at the Gryffindor tower entrance, forever adoring good food, drink, and the highest security, long after the Fat Lady herself had passed away into wizarding history.

The idea of magical portraits was ingeniously altered to suit the plot. In the *Goblet of Fire*, which begins with gossip in a local tavern, the people in the portraits do indeed move from frame to frame, playing Chinese whispers with the latest chinwag. At more festive times, when the wine flowed and the living was easy, the subjects of the portraits got a little plastered. And, after Hogwarts had a spring clean, the portraits complained about the cleaning and grumped about their skin feeling a little raw. The magical portraits become beacons of ominous portent too. In the *Prisoner of Azkaban*, the Fat Lady became a plot focus when knife slashes were found in her frame.

And yet magical portraits had certain limits in space and time. Few portraits were capable of an in-depth analysis of the more intricate aspects of their lives. The portraits really were mere two-dimensional depictions of the sitting wizard or witch, as seen by the artist. And yet the rare magical portraits were capable of so much more. They could countenance far more interaction with ongoing events in the living world.

Consider, for example, the portraits at Hogwarts of headmasters, or headmistresses, painted before their death. Once the portrait was done, the head teacher in question could store the portrait away and regularly visit it, instructing the interactive painting how to act and behave just like himself or herself. In so doing, the head teachers transferred many insights, much knowledge, and useful memories that could be useful to future successors of the same office. The range and depth of wisdom held within the headmasters' office was thus profound. Those who accepted the surface impression of the office as sleepy and inactive really were missing the vital point of the very presence of the magical portraits.

But what progress has been made towards moving portraits in the muggle world? Or is Picasso correct when he suggests we've essentially discovered nothing in the last 30,000 years since the days of prehistoric cave portraits?

The Muggle Moving Portrait

Imagine some of the greatest moments captured in a muggle moving portrait. The greatest soccer goal of all time, by Diego Maradona, for example. A truly fabulous few seconds from the 1986 World Cup as Maradona pirouettes through 180 degrees, slips deftly between player after player, slaloms deep into the penalty box, dummies the keeper, and adroitly slips the ball over the line and into a billowing net.

Or perhaps a moving portrait of one of the greatest ever works of art—*The Garden of Earthly Delights*. A triptych painted by Dutch master Hieronymus Bosch between 1490 and 1510, the painting is among the most intricate and enigmatic paintings of Western history, and filled with iconography and symbolism that have sparked debate for centuries.

Bosch's masterpiece is a mysterious invented world, full of strange and daunting details. This imposing portrait features a man with a tree for a body, who gazes out from Hell, giant birds dropping fruit into the mouths of naked people, slithering creatures invading paradise, and a devil-bird that devours a man whole. What is the meaning of all this, the most famous of paintings? Perhaps Heaven and Hell are not the destinations of your soul, but states of being that live inside you—no one knows for sure. But a muggle moving portrait of *The Garden of Earthly Delights* could be interrogated about its meaning.

And then we could develop moving portraits for famous moments from history. The self-immolation of Buddhist monk Thich Quang Duc in 1963, perhaps, who set himself on fire in protest of the persecution of Buddhists by the South Vietnamese government. Or maybe the footprint of astronaut Neil Armstrong on the surface of the moon. A feat that would have derived scorn only decades before, the historical achievement of all human nations in space meant that, no matter what happens to humans in the future of this planet, his footprint will remain.

But how would a muggle moving portrait be made? There is, of course, the GIF. GIFs have wobbled across thousands of webpages, fluttered within myriads of Facebook profiles, and transformed countless embedded Tumblrs. GIFs can be seen in animated advertising, the sign-off signatures of email, and social media avatars. In short, GIFs are everywhere. The acronym 'GIF' stands for 'graphics interchange format'. The image format was designed for a digital space that was just coming of age. Developed by Steve Wilhite of Compuserve in June 1987, the GIF began as black and white image transfers, then moved to 256 colors, while still keeping a compressed format that the slow internet speeds of the day could easily handle. Today, it seems, folk are fascinated by the GIF, as it has become de rigueur on the web, a default brand of internet humor, and an essential for viral YouTube videos.

But can the GIF transfer to muggle newspapers, like the magical portraits in wizarding world newspapers such as *the Daily Prophet* and *The New York Ghost*? Britain's *Empire* magazine claims to have taken inspiration from the wizarding world and produced the world's first moving-image cover. The magazine published the image in a limited-edition celebration of the launch of the Potter spin-off movie, *Fantastic Beasts and Where to Find Them*. The limited-edition cover looked something like an enchanted newspaper, and was modelled on *The New York Ghost* newspaper from the *Fantastic Beasts* story.

The *Empire* images move, with two such portraits embedded into the magazine cover. The tech hidden in the magazine cover (a double layer of card and an embedded video-screen) allowed reader interaction by enabling a press play option on the portraits. Beneath the card sat the necessary microchips and circuit boards, enabling the portraits to come to life with the press of a button. The portraits in question were an exclusive behind-the-scenes clip of *Fantastic Beasts*, and another showing the movie's trailer. It may not yet be an artificially intelligent muggle version of a magical portrait, but it's a start.

HOW COULD YOU MAKE YOUR OWN MARAUDER'S MAP TO SKIP CLASS?

Picture yourself on a wet Wednesday afternoon. You're trapped in double business studies, the most tedious topic known to any school syllabus. No need to panic. A cunning plan is afoot. And it involves a magic map. Outside your tutorial of torture lies a Byzantine network of classrooms and corridors. Your mission, should you choose to accept it, is to negotiate this Kafkaesque chaos and escape into the life-giving sunshine, beyond the school sentries and boundary surveillance. But wait, what exactly does this magic map do?

In the Harry Potter Universe, such a magical document was known as the Marauder's Map. With this map, the intricate layout of the sometimes subterranean, seven-storied, one-hundred-and-forty-two staircased, towered, turreted, and deep-dungeoned architecture of Hogwarts School of Witchcraft and Wizardry was clearly conjured up.

The map was an all-seeing eye into the medieval castle's deep, dark heart. The map spied each and every classroom, all hallways, and every creepy castle corner. The castle grounds also fell under the map's purview, as well as all the clandestine corridors hidden within its walls. Nor did witches and wizards escape its knowing reach. Each one was signified on the map by an animated set of footprints and a scroll-like caption. The Marauder's Map was not fooled by Harry's invisibility cloak, Animagi, or Polyjuice Potions. Even the Hogwarts ghosts were captured by its gaze.

True, the map was not infallible. It could not distinguish wizards or witches bearing the same name, for example. Neither does the map show unplottable rooms. The Room of Requirement, for example, was revealed by Dobby the house elf, and not the map itself, which seemed not to even know the Room existed. And the same was true of the Chamber of Secrets. It never appeared on the map. As with the Room of Requirement, the Chamber may not have been shown simply because the map's creators, Remus Lupin, Peter Pettigrew, Sirius Black, and James Potter—also known as Messrs Moony, Wormtail, Padfoot, and Prongs, "Purveyors of Aids to Magical Mischief-Makers" —simply never knew of its existence. So, what would we need to make a muggle version of the Marauder's Map?

Medieval Map Makers

Like Hogwarts itself, the golden age of map making was medieval, and it began with ships. Two Chinese inventions, the compass and the sternpost rudder, had a global effect at sea. Long voyages became viable. The seas were thrown open to exploration, piracy, a colossal expansion in trade, and war. The need for better navigation had profound consequences for map making. An open ocean meant more accuracy: better observations, better instruments, and better maps. So open-sea navigation raised the need for a brand new quantitative geography, and the desire for devices that could be used onboard ships, as well as on land. And so the obsession with longitude began.

The great European sea voyages started around 1415, and opened up the planet to plunder. The voyages were the fruit of the first conscious use of geography in the pay of glory and profit. Fledging empires soon realized they were able to exert global control based on knowledge of territory: knowing *where* you were and knowing *what* you owned. And so, navigation and mapping became even more important to trade. But the golden age of maps led to a golden age of piracy.

The pirates that preyed upon the high seas, an echo of rival trade and colonization attempts by European powers, often sought a surprising booty. If a sea raid proved successful, the boarding pirates would head straight for the hold. Rather than gold, silver, or pieces of eight, the most precious cargo a ship possessed was its maps and chronometer. Indeed, some cartographers would knowingly include errors on their maps, to mislead the uninitiated should the map get into the hands of the wrong kind of pirate.

Such cyphered anti-pirate maps bear a resemblance to the Marauder's Map. The Marauder's Map was also coded, normally disguised as a blank piece of parchment. To view the map, a wizard or witch had to tap it with their wand and say, "I solemnly swear that I am up to no good." Only then did the map reveal itself. Similarly, to once more hide the contents of the map so the parchment again appeared blank, a wizard would tap it and say, "Mischief managed." The only difference was this. Medieval maps were meant to guard against mischief, but the Marauder's Map was designed specifically to *cause* it!

Tagging Teacher

A muggle version of the Marauder's Map might be based on GPS. GPS, or the Global Positioning System, is a network of around thirty satellites, which orbit the Earth at a height of 20,000 kilometers. As with much technology before and after it, the system was developed for the military, the US military in this particular case. But eventually, anyone with a GPS device was allowed to use it. That's whether you have a mobile phone or a plain GPS unit and can receive the radio signals, which are broadcast by the satellites.

No matter where you are on Earth, GPS will find you. Wherever you may roam, at least four GPS satellites are visible. Each of the satellites transmits data about *where* it is, and *when* it is. These data signals, beaming down at the speed of light, are picked up by your GPS. Once such data is done for at least three satellites, your GPS knows where you are. Your GPS receiver does this by a process known as trilateration. Now, imagine applying the technique to a school escape situation. Imagine a teacher is lurking somewhere within the school catacombs. High in the sky above sit the beady eyes of three satellites. Let's call them satellites A, B, and C. If the lurking teacher is spied by satellite A, then that satellite will know just how far away he is. And if satellites B and C also spy the teacher, they too will read his position. So, taking all three readings together, where they intersect is the exact spot of the lurking teacher. And the more satellites there are above the horizon, the more exact will be the reading of the teacher's position. All of this is done with a little help from Einstein.

To ensure the very best in time accuracy, GPS satellites carry atomic clocks. Einstein's theories of Special and General Relativity predicted that an atomic clock in Earth's orbit would show a slightly different time to an identical clock down on Earth. Einstein's brilliant brain realized that time runs slower under a stronger gravity. So, the clocks on board the satellites will seem to run faster than their Earth-bound counterparts.

The satellites must make a correction for speed, as well as gravity. Each satellite in the GPS constellation orbits at a height of around 20,000 kilometers. And at that altitude, they speed along at about 14,000 kilometers every hour (that's an orbital time of roughly 12 hours—contrary to popular belief, GPS satellites are not in geosynchronous or geostationary orbits). And as they travel at such speed, Einstein's Special Relativity predicts the satellites' clocks will appear to run more slowly than a clock on Earth! So, the whole GPS network must make allowances for these relativistic effects of gravity and speed on time.

Yet none of this should worry our potential escapee because the tech for tagging teacher is already with us. GPS security tags, used for tracking pets, people, or even prying teachers, are already on the market, are tiny, and are solar-powered. They are accurate to the meter, and work indoors. Using a super sensitive patch antenna, and being smaller than two flat AA batteries, these GPS security tags use a GPS satellite constellation. And all this means that, armed with a smartphone, a school escapee can easily plot current and previous positions of teachers simply by using a mapping app. So remains one last challenge: actually getting the tag *onto* teacher . . .

HOW WOULD WE FASHION
A WORKING WEASLEY
FAMILY CLOCK?

People have a habit of simply vanishing. Famous Chinese sage Laozi mysteriously disappeared. He vanished in 531 BC, leaving behind his famous book, the *Tao Te Ching*, the basic teaching of Taoism. Spartacus, leader of the slave rebellion against the Roman Republic, also departed in a proverbial puff of smoke in 71 BC. And Amelia Earhart, the renowned American aviator, disappeared on July 2, 1937, after becoming the first woman to try a circumnavigational flight of the globe.

But vanishing unexpectedly and without explanation was never a problem for the Weasleys. And that's because they had the famous Weasley clock. This was no ordinary time device. Rather than dully denoting the time of day, the Weasley clock instead monitored the whereabouts of individual family members. Located in the living room of The Burrow, the Weasley family home on the outskirts of Ottery St. Catchpole in Devon, England, the clock boasted nine golden hands, one hand for each member of the household.

The hands told the tale of where each Weasley was. Set out on the clock's face was a series of optional locations, including School, Work, Traveling, Home, Lost, Hospital, Prison, and even Mortal Peril. In addition, were the more jocular categories of "time to make tea" (this *is* Britain, of course), "time to feed the chickens," and "you're late." Each family member's whereabouts could be seen at a glance, as their assigned clock hand would be pointing to wherever on earth they were. Given his infatuation with all things muggle, was the head of the household, Arthur Weasley, ever tempted to manufacture something similar for the non-magic market?

And, if such a business venture was a going concern, what tech might be used to do the job?

Wizard Clocks

Believe it or not, wizardy muggle clocks have already been built. So, let's go through such a build. First up, an overview. As with the Weasley clock, our wizard clock will have clock hands to represent people, and clock "times" to represent locations. Our wizard clock will also give an at-a-glance guide to the whereabouts of family or friends. And the kind of magic our clock will use will be a muggle tech that runs from smartphone to webserver to wizard clock. You'll have enough of an idea of where your loved ones are, so you will either be able to stalk them for fun, or know when it's best not to call.

Surely, the dream wizard clock of choice is a grandfather clock, set in a wooden frame. Since the clock has to show people and places, we need to add to the typical two- or three-handed clock to make as many hands as we need. But, for now, let's assume we make a grand total of four hands, one hand for each of the four houses of Hogwarts: a Professor Dumbledore hand to represent Gryffindor, a luminous Luna Lovegood hand for Ravenclaw, a Professor Pomona Sprout hand for Hufflepuff, and finally a Professor Snape hand for Slytherin.

Our wizard clock-face should also show the most commonly used locations. Let's pick some appropriate places for each of our four hands. As Dumbledore has a keen interest in the heavens with his own telescope in the Headmaster's Office, let's have a first location of "Astronomy Tower." The wonderfully dotty witch, Luna Lovegood, lives in a rook-like house in Ottery St. Catchpole, so let's name our second location "Ottery." Pomona Sprout is often found pottering around in the Hogwarts herbology green-houses, so our third Clock location will be "Greenhouses." And finally, Severus Snape is a very talented Potions Master, so our last clock location shall be "Dungeons," where he synthesizes such elixirs. In short, we have four enticingly named clock locations of Astronomy Tower, Ottery, Greenhouses, and Dungeons! To these we might also add the relatively normal locations of Traveling, Lost, and possibly even Mortal Peril.

A Weasley clock managed its own magic, but our wizard clock needs muggle tech. Dumbledore, Lovegood, Sprout, and Snape, who together sound like the most Dickensian set of solicitors in existence, would each need to carry some kind of mobile tech. This could take the form of a smartphone, upon which an appropriate Android or iOS app had been written and which held the same core functionality. The Android app would run a service in the background, updating its location at timed intervals. In contrast, the iOS phone app would send its location whenever the device talked to a new cell tower, always assuming they had such towers at Hogwarts. With the Android, location data is up-to-date and accurate. With the iOS phone, super-accurate and super-frequent updates can be sent by simply keeping the app open and running. Both apps also support sending destined locations, either for testing or just to clarify a user's whereabouts when their phone's location isn't as accurate as it might be.

Mortal Peril

And so, wherever the user roams, a signal of where they are in the world is sent. The whereabouts would wing their way to a webserver; a computer system, or program, that beams data to users. In this case, the intel would be sent to our wizard clock. Once received by the clock's on-board Wi-Fi module, the clock has four servomotors that enable it to travel through a full range of potential locations. And so, like the Weasley clock that inspired it, our Wizard clock has done its deed: the whereabouts of Dumbledore, Lovegood, Sprout, and Snape can be seen at a glance, as our clock hands will show exactly where they are.

The real test of our wizard clock is the question of Mortal Peril. How on earth are we going to cope with that, using only muggle tech? Perhaps the best solution is using wearable tech. A smartwatch, for example, can simply be furnished with an app which, in real-time, is able to detect the stress associated with the body's fight-or-flight response. The app could pick up signs of whether the witch or wizard wearer was experiencing changed levels of heart rate, perspiration, blood pressure, or movement. And so, our plan for a real-life wizard clock is done. Don't just sit there; go and build one!

CAN TECHNOLOGY REPLICATE THE REDUCTOR CURSE?

T he reductor curse is used to destroy solid objects by splitting them into pieces or a fine dust. Harry used it against a hedge but only managed to burn a small hole in it, while it has also been used to explode shelves.

If we wanted to achieve the same effect, we would have to identify the solid material to be destroyed and then find the most suitable method to do so. In essence though, whatever method we use boils down to finding a way to break the bonds that hold the material together. This would generally involve some kind of physical, chemical, or biological reaction. So, in what ways could we replicate the effects of the reductor curse?

Chemical Reactions

A chemical reaction is one where electrons are gained or lost by a substance. If no electrons have been traded, then it isn't regarded as a chemical reaction. Common chemical reactions include milk going sour, rusting of iron, and combustion, i.e., burning. In all chemical reactions, chemical bonds between atoms and molecules are broken and new ones are formed.

What chemical changes could break down a solid into a fine dust or else cause it to burn, explode, or split into pieces?

To burn a small hole in a hedge would generally imply that combustion has taken place. A combustion reaction causes a substance, called the oxidizer, to react with another substance, called the fuel. The oxidizer acts to take electrons from the fuel and releases energy in the process. The fuel is then considered oxidized. There are different oxidizers with different strengths, but the major oxidizer of combustion reactions on Earth is oxygen, which is why things on Earth generally need oxygen to burn.

In a combustion reaction, oxygen takes electrons from the fuel, and we say that the fuel has been oxidized because it has lost electrons. The process by which oxygen *gains* electrons is called reduction. While the fuel is oxidized by oxygen, the oxygen is reduced by the fuel. In fact, you can't get one process without the other.

Whenever something has undergone oxidation i.e. been oxidized by having electrons taken, the substance that has taken the electrons (the oxidizer) must have undergone reduction i.e. it has gained the electrons. As such, these types of reactions are also known as redox reactions, which is shorthand for **red**uction-**ox**idation reactions. Combustion is a fast-acting redox reaction, whereas a corrosive process like rusting is a much slower redox reaction.

It may be tempting to relate chemical reduction to the reductor curse, but in the case of burning a hedge, the wizard would be more correctly oxidizing the hedge through combustion. Could the reductor curse be an extreme form of oxidation, then?

Extreme Oxidation

There are various oxidizers with different abilities to pull away electrons. One of the strongest is an extremely reactive and highly toxic substance called chlorine trifluoride. Chlorine trifluoride is hypergolic, meaning it spontaneously ignites when mixed with other substances. It readily reacts with all known fuels, as well as cloth, wood, asbestos, sand, people and water, with which it reacts explosively. It was investigated for use as a possible rocket fuel but subsequently deemed too dangerous.

If a wizard's reducto spell instigated a chemical reaction like oxidation of a wooden table, it would still take a fair bit of time to actually burn the table down to an ash or fine dust. The wizard would have to manipulate the speed of the reaction to make it happen quicker. In chemistry, catalysts can be used to achieve this, but increased temperature can also help. A catalyst is a substance that speeds up the rate of a reaction, without being used up in the process or chemically changing itself.

When Parvati Patil "produced such a good reductor curse that she had reduced the table carrying all the Sneakoscopes to dust," perhaps her spell rapidly oxidized the wood, using some kind of magical super powerful catalyst. However, as muggles have discovered, rapid oxidation could be a very dangerous thing to attempt indeed, considering the explosive reactions that can occur. Other than chemistry, what other means could we use to imitate a reductor curse?

Physical Reactions

A physical change (reaction) is one where there is a change in some aspect of a substance (such as temperature, shape, color, size) but no change in the composition of the substance. For example, when ice is heated it changes from a solid to a liquid but still has the composition of water. The water is said to have changed state in this case i.e. from solid to liquid.

Other examples of physical changes include crushing a can, boiling water, breaking a glass, demolishing a building, and grinding peppercorns. If a reductor curse acts by causing a purely physical change, then, depending on the target, there are a few physical processes that could create that effect.

A shockwave from an exploding rocket or meteor in midair can smash windows, though the solid concrete parts of the building would remain comparatively unharmed. For a shockwave to disintegrate or pulverize a wall or rock face would require an immense amount of energy. There is still a way to break up rock, though, and it doesn't use explosives.

Things in nature are subject to the elements such as wind, rain, and heating by the sun. The elements' effect on natural objects is known as weathering. This is frequently seen when water gets into cracks in certain rocks and repeatedly freezes and thaws. The water expands and contracts over and over again, developing stresses in the rocks. Eventually, blocks fracture and break off, ending up as a scree slope at the base of the rock face.

Weathering is a rather long-winded process and would only be possible on certain types of rock; it wouldn't work on wood or a hedge as they are more flexible and can withstand the stresses and strains. So, how about we cool things down a lot more.

Frozen Solid

When some objects like polymers, flowers, or fruit are frozen to extremely low temperatures they can be simply crushed or else shattered with sufficient impact. This works because as the temperature decreases the material becomes more brittle. Liquid nitrogen is a common substance used t⌐ bring the temperature down, as it typically only exists as a liquid between minus 210 and minus 196 degrees Celsius.

At room temperature, the structure of a substance can better absorb an impact, dissipating any stress and strain by stretching and deforming. At this temperature, the molecules in the structure are free to slip past each other. However, as the temperature is reduced, the material becomes less elastic until it reaches its frozen state. At that point, the molecules are less free to move and so the energy of the impact is not dissipated but rather concentrated in localized regions, leading to brittle fractures and potential shattering.

Brittle fractures are a result of the breaking of atomic bonds, which lead to the substance being cleaved apart on a molecular level. The energy put into causing the break is required to overcome the cohesive forces between the atoms along the crack path. We see this when wood is broken into chunks and giving off splinters, but it wouldn't turn the wood into dust. Usually, this requires repetitive action from a saw or chopper.

If wood or paper is frozen, they don't become brittle in the same way because they are made of fibers which, although may become brittle in themselves, can still slide passed each other and so allows a level of flexibility in the object as a whole. So, reducto couldn't be used on a table by making it extremely cold; the table would break but not shatter or be reduced to dust. How about using sound to break up a solid object?

Sonic Reducto

The use of sound waves to break up solid materials is a well-known phenomenon, as seen in the ability of some singers to shatter glass using their voice alone. Muggles have a contraption called a lithotripter that can use sound to pulverize solid masses in the body known as calculi but commonly called stones. Calculi include bladder stones, gallstones, and kidney stones. They form as a result of minerals and salts clumping together after becoming highly concentrated in the urine. The stones are usually quite small and get passed out unnoticed in the urine. However, if they are too big, then treatment to remove them may be necessary. The lithotripter allows provision of a noninvasive treatment such as ESWL, which stands for extracorporeal shock wave lithotripsy. Extracorporeal simply means "outside the body."

In ESWL, the lithotripter is used to focus high-energy ultrasound shockwaves at the stone. The shockwaves travel through the body alternately compressing and stretching tissue as they pass. Most body tissue is quite resilient to the resulting tensile forces, but solid materials such as the stones have less give and are more susceptible to fracturing. For most of the journey through the body, the waves are spread out so they don't impart as much energy. However, when they reach the focal point where the stone is, the energy from the shockwaves are more intense and sufficient to cause the stone to break up into smaller pieces, some as small as sand grains. It should be noted that it requires more than a thousand shockwaves to pulverize the stones, with some treatments lasting up to an hour. Again, the rate at which the breaking up occurs is still too slow for the immediate needs of the brawling witch or wizard. So where does this leave us?

Reducto!

Burning a hole in a hedge is a very possible technology. Blowing up glass objects is also possible either by using sound waves that match the natural frequency of a glass or more broadly through the creation of shock waves. These could smash glass shelves if carrying enough energy, however they would need to be many times stronger to break brick.

On a smaller scale, it's also possible to break down internal stones, or calculi, using supersonic sound waves through the body. Wood is more difficult to reduce to dust, unless you're willing to wait a number of minutes for it to burn down, or else find some physical object to break it up. So, we do have technologies that can replicate aspects of the reductor curse, but a fully functioning one-size-fits-all technology is currently out of the question.

HOW CAN A WIZARD MAKE GREAT BALLS OF FIRE?

Fire. The holy grail for early humans. Prometheus was said to steal it from the gods to benefit mankind, but however we came across it, once we discovered how to make it, the world changed. We went from maintaining naturally occurring fires caused by things like lightning strikes or volcanic activity, to working out how to rub sticks together or strike a rock to ignite some tinder. Now, we can just use matches or grab a lighter, but to ancient humans, these devices would appear as magical as someone proclaiming "*Incendio*" or "*lacarnum inflamare*" to send flames flying from the tip of a magically modified stick. So, how could a wizard create a fireball from a wand?

A Recipe for Combustion

Fire is the result of a chemical reaction called combustion. In combustion reactions, substances react together, resulting in the production of new substances. Heat and light are given off in the process. A reaction that gives off heat is called an exothermic reaction. Therefore, combustion is exothermic.

In the majority of combustion reactions, an oxidizer reacts with a substance (the fuel) when presented with enough thermal energy, i.e., heat. An oxidizer is basically a substance that takes electrons from the substance it is reacting with, i.e., the fuel. On Earth, the most common oxidizer in combustion reactions is oxygen, as it's widely available, making up 21 percent of the air we breathe. Other possible oxidizers that can lead to combustion include fluorine or chlorine gas.

The creation of fire on Earth generally requires the presence of three ingredients; fuel, heat, and oxygen. These are commonly referred to as the fire triangle. If any one of these elements is lacking, combustion won't occur, and no fire can be made. So, for a wizard to successfully conjure a flame from the end of their wand, all three parts of the fire triangle must be present. Since Harry Potter takes place on Earth, the oxygen part is covered. However, if in a place with very little or no free oxygen, the wand would need to produce its own oxidizer. This is how rockets work in space. The fuel they use contains its own oxidizer, allowing rockets to burn in the vacuum of space even though there's no oxygen. For our purposes, we'll only be considering environments where there is sufficient oxygen, so the trick is in working out how to create combustion with a wand.

Fire Wands

A wand is seen as a way for a witch or wizard to channel their magical prowess. Each one is made from a particular type of wood and contains a magical core that affects how the wand behaves. But how could a wand manipulate oxygen, fuel, and heat to create fire?

In the muggle world, lighters are a common way to make a flame and have been around in various forms for more than a century. They work by igniting a fuel in the presence of oxygen. It's not hard to make a lighter in the shape of a wand; BBQ lighters can be especially long already, and products described as wand lighters actually exist. Of course, a wand can do so much more than just make a flame, so filling a wand with a lighter fluid such as butane or naphtha wouldn't leave much room for other functions. It's a start, though.

If you take a transparent disposable lighter, you can see that it contains liquid fuel. This liquid is just a gas under lots of pressure. Generally, if gases are compressed enough they will become a liquid. A gas can also become a liquid if it's cooled sufficiently, and if it's cooled even more, it can freeze into a solid. A fuel will take up less space as a liquid or solid and so any fuel source held in a wand would be stored more efficiently in one of those states.

We're surrounded by fuels that are in solid, liquid, or gas states. Solid fuels include coal and wood, but they don't actually combust straight into a flame. When they get hot enough, they undergo a reaction called endothermic pyrolysis, which produces flammable gases. It's these gases that can then combust to produce light and heat. This heat feeds back into the process as part of a chain reaction, which is another important part of fire propagation. So much so that it's now often included with the three elements of the fire triangle to form what's referred to as the fire tetrahedron. So, how could a wand achieve this?

Considering that the outside of a wand is made of wood, there's always the possibility that it could set alight. Having said that, combusting a bit of the wand itself could technically be a solution to fueling a fire, but this would vastly limit the amount of fireballs that could be produced before the wand was burnt up, leaving a charred black stick that would be brittle as well as messy. This is clearly not an option, so what fuels could a wand conjure to make fire?

Fueling the Fire

When solid fuels are heated, they undergo changes that release combustible gas. Liquid fuels work in a similar way in that when they're heated they evaporate into a flammable gas that is then combusted. So really, regardless of the initial state, combustion generally occurs using flammable gases, like the fuel in a fire triangle or tetrahedron.

Among other things such as available oxygen and air pressure, the particular fuel being burnt can have an effect on the temperature of the resulting flames. This is important because some objects need more heat to be set alight. So, a lower temperature flame from something like animal fat or kerosene would need to be bigger or applied longer than a higher temperature flame from a fuel mixture like oxyacetylene.

Muggle magicians often make fireballs, and they have different methods to produce them. One example uses a flammable solid material that's basically cotton wool but supercharged. It's called nitrocellulose or flash cotton, and when it's burnt it produces a flame that is literally gone in a flash. The diffuse nature of cotton wool allows oxygen to reach the sites of combustion more easily, making the reaction happen a lot quicker. There's a small product out on the market called Pyro Mini Fireshooter that launches fireballs in this way, but from a smaller unit. A problem with this technique is that it burns so quickly that it wouldn't pass a great deal of heat to whatever it comes into contact with. It's the reason why show people are so content with setting light to their bare palms. It also means that Hermione couldn't use it to set light to Snape's cloak. So, what other methods are there?

The spraying or ejection of liquid fuels could also be employed. A stream of flammable liquid can carry a flame a lot further than a gas fuel can. These liquid fuels can also more readily land on objects and continue burning, especially if they are thickened; such was the case with napalm. Most flamethrowers used in World War II and the Vietnam War operated in this manner.

If a wand was loaded with a bit of liquid fuel, it would also need a way to propel it. It could use a pressurized gas as the propellant (which is how flamethrowers do it) or be mechanically squirted out by squeezing the base of the wand. Both of these methods would work, but the resulting flame would look more like a stream of flame rather than a fireball, which leaves gases, of which methods are various.

A wand could contain compressed gas fuels as liquids. This wouldn't take up much space and could provide a few good burns for a small fuel payload. Reactions could also happen within the wand to create gases. One such example is calcium carbide, which reacts with water to produce the flammable acetylene gas. In this way, the wand just needs a chamber to contain the calcium carbide, and the ability to add water to it at will. The pressure of the gas could build up and be released as a puff, which once ignited could provide a small fireball. Regardless of the fuel type and method of release, all of these methods rely on a sufficient source of heat for ignition, so how could that be done?

Generating Heat

Lighters have various methods of ignition to provide the necessary heat energy for combustion of the fuel and oxygen mixture. The most common method uses a spark from a rock or metal such as ferrocerium i.e. lighter flint. The drawback with the flint is that it requires mechanical action to create the spark, which would have to somehow happen within the wand. The same is true with using an electric discharge from a squeezed piezo electric crystal; something found in many lighters. In this type of lighter, an electric arc is generated between two electrodes when sufficient voltage is applied.

There are also catalytic lighters which use an alcohol like methanol in the presence of a platinum catalyst. A catalyst is a substance that allows a reaction to happen faster or with a lower input of energy, but isn't used up itself in the process. As the methanol vapor comes into contact with the platinum, a chemical reaction occurs, which generates heat. This heat is enough to initiate combustion in the methanol. This would be a compact and easy enough way to fit a lighting mechanism onto the tip of the wand, and a pressurized gas again would allow a flame to be propelled from the tip rather than just emanating from it.

Another possible option is to use what's called hypergolic fuels. With these, two substances come into contact with each other and undergo a reaction that causes them to combust without the need for a separate ignition source. These are frequently used in space flight. As far back as the Apollo missions, the fuels Aerozine 50 and nitrogen tetroxide were used in the engines of the service and lunar modules.

Great Balls of Fire

So, it *is* possible to create a fireball, providing all elements of the fire triangle are present. Fitting these within a wand is the issue. Using gas as a fuel could produce a nice fireball as could any reaction that produces a significant volume of flammable gas. The fuels could only be stored in small quantities, though. To provide the heat for the reaction, hypergolic fuels are possibly the most straightforward method, but ignition via catalysis also seems like a good way to achieve a flame on the tip of a wand. Either way, a fire-producing wand is well within the reach of real science and technology.

PART III
HERBOLOGY, ZOOLOGY, AND POTIONS

IS THE BEZOAR A REAL ANTIDOTE?

There are plenty of muggle drinks that can render the supper senseless. Absinthe has been known to induce hallucinating effects, Bruichladdich is an incredibly pure and potent whiskey, and Spirytus Rektyfikowany is a Polish vodka that makes you meet your God when you overindulge. But the tipple that almost did in Ron Weasley was a plain old oak-matured mead. Poisoned. It was Ron's glass that hit the floor first. Ron followed. He crumpled to his knees, tumbled onto a rug, spasmed spookily, and as foam oozed out of his mouth, his skin turned blue.

Harry to the rescue. Looking about and leaping up, Harry hurriedly stripped the walls of its potions. A box fell forward and out spilled a scattering of stones, each no bigger than a bird's egg. Harry took one of the dry and shriveled stones, opened Ron's jaw, and thrust it into his throat. At once, Ron stopped moving. But soon, a great hiccupping cough, and Ron was back. Breathing. The quick-thinking Harry had used a bezoar.

The bezoar is an undigested clump of matter, taken from the gut of a goat. Such clumps accumulate inside digestive systems, and are usually made of hair, fibrous plants, and are similar to a cat hairball. In the magic world, bezoars act as antidotes to most poisons, with Basilisk venom being an important exception. But, as bezoars are also real-life objects, what exactly *are* they, and what can they actually do?

Bezoars

Bezoars were believed for many centuries to be the most marvelous medicine. The word bezoar (pronounced bē zōr) is thought to derive from the Persian *pâdzahr*, which literally means antidote, or counterpoison. Now, there are over two million species of animals on Earth, but only 148 species are suitable for farming. Of those 148, only 14 have ever been successfully farmed: goats, sheep, pigs, cows, horses, donkeys, Bactrian camels, Arabian camels, water buffalo, llamas, reindeer, yaks, mithuns, and Bali cattle. Just 14 large animals in over 10,000 years of farming.

Of these 14, the 4 big livestock animals—cows, pigs, sheep and goats— were all native to what we now know as the Middle East. The area that was home to the best-irrigated crops in the world was also home to the best-watered animals. Little wonder that this area became known as the Fertile Crescent, and little wonder that this area is also the home to the discovery of the bezoar.

Science took a huge leap forward in medieval Islam between the 8th and 14th centuries. Its legacy is still with us in the forms of algebra, algorithm, and alkali. All are Arabic in origin and sit at the very heart of contemporary science. During these centuries, Islam was a diverse, outward-looking culture. The people were charmed by knowledge and quite fascinated by questions of science. The astronomer and mathematician Al-Biruni assessed the size of the Earth to within a few hundred miles. The physicist Ibn al-Haytham helped found the science of optics. And Islamic scholars, obsessed with precise measurement in fields such as astronomy, had a great impact on the scientific revolution that took place in 16th and 17th century Europe and helped shape the work of people like Copernicus.

As far back as the 7th century, the Islamic world used bezoars. They were usually hailed from goats, but also from the guts of deer, camels, cows, and other ruminants. Before being administered, the bezoar would be pummeled and ground into a powder, and either gulped down, or taken with hot water in the form of a more civilized tea. The belief in the benefits of the bezoar was so great that it was also made into a bandage or poultice. In this form, it could be taken externally as a remedy for fever, epilepsy, or even leprosy.

While Islam blossomed, Europe was going through its Dark Ages. In the 14th century, between 25 percent and 60 percent of the European popula- tion, an estimated 50 million souls, are thought to have perished in the Black Death, the bubonic plague that swept through Europe and parts of Asia and Africa. The idea of the bezoar, a potent and powerful medicine, must have appealed greatly to the suffering Europeans. Indeed, King Edward IV of England is said to have attributed his recovery from a pustulent wound to his doctor's use of a bezoar. The celebrated Islamic physician, Avenzoar, is said to have been the first to write for Europeans on the bezoar.

As news of its healing properties spread, bezoars became more commonplace on the continent. However, medical research does not seem to have been Napoleon's forte. When the Emperor of Persia gave several bezoars to the Frenchman, the story goes he tossed them away before his death, which could have been down to poisoning!

Bezoars as Gems

Napoleon aside, the fame and fortune of bezoars skyrocketed. They soon appeared among lists of gems. A price list drawn up by a Germanic apothecary in 1757 offered sapphires and emeralds, rubies and other precious gems, some of which were for medical means, but the real pick of the precious list was the bezoar. Its listed value placed its price a clear fifty times the estimate for emeralds.

Bezoars would be worn as amulets, and as amulets were meant to contain within them some quality or power to protect their owner from harm, bezoars were a perfect fit. They would be worn around the neck or carried in boxes jeweled to the hilt. Queen Elizabeth I of England, whose eponymous reign and era is associated with the likes of Shakespeare and Marlowe, had several bezoars set in rings. They later became part of the crown jewels of the realm.

A cottage industry sprang up in bogus bezoars. An English goldsmith was summoned to the courts for allegedly peddling worthless fakes in the early 17th century. Hardly surprising, as the asking price of one of the fakes was a cool 100 pounds, worth around $40,000 today. Around a hundred years later, in 1714, bezoar issues were raised by a Fellow of the Royal College of Surgeons in London. Local drug suppliers alleged they had five hundred ounces of bezoar in store. The surgeon had begun to smell a rat when he calculated that such a stock of bezoars would necessitate the slaughter of around fifty thousand goats.

As to the efficacy of the bezoar as an antidote, consider this tale. King Charles IX was gifted a bezoar. Seeming as skeptical as his countryman Napoleon, the king called up his royal physician, Ambroise Paré, wishing to know whether the stone really did have the power to protect against all poisons. "Nonsense," replied Paré, as no two toxins are exactly the same, no single stone could have the substance to be a universal antidote. "Fine," said the king. "Then we shall test it to find the truth." So, the king called up a convicted criminal who had been sentenced to hanging and was soon due to die. A new choice was presented to him. Eat a deadly poison, and then the bezoar. If he was cured, he could go free. The condemned man was no skeptic. He downed a poison prepared by the royal apothecary, greedily followed by gulping down the bezoar. Lo and behold, he died in agony a few hours later, followed by the sizzling sound of a stone as the king tossed the cure-all into the fire.

DEVIL'S SNARE: WHAT ARE THE REAL-LIFE FLESH-EATING PLANTS?

Herbology is one of the many fascinating topics that students study at Hogwarts. Herbology is all about magical plants and fungi, and although it is sometimes overlooked, Professor Sprout's lessons help Harry, Ron, and Hermione out of many sticky situations during their time at the Castle.

One of the deadliest plants that Harry encounters, both in and outside the herbology classroom, is Devil's snare. In the golden trio's Voldemort-busting adventure through the trapdoor, Devil's snare is one of many trials that they have to overcome to reach the *Sorcerer's Stone*. This particular magical plant has the startling ability to constrict around its prey, and as Ron and Harry discover, the more you struggle, the tighter it squeezes. Later in the series, Devil's snare is featured as a deadly weapon more than once. Neville Longbottom and Professor Sprout position them strategically around the grounds during the Battle of Hogwarts to take down the giants and death eaters invading the Castle.

In another sinister scene, Devil's snare is smuggled as a method of assassination into St. Mungo's Hospital for Magical Maladies and Injuries. A potted Devil's snare was delivered to an unsuspecting comatose patient, Broderick Bode, and mistaken for an innocuous Christmas gift. The death eater responsible, Walden Macnair, was able to sneak the plant passed the Healers, where it strangled Bode to death before anyone could realize.

Luckily for Harry and Ron, there is an easy way to escape Devil's snare. The harder you struggle against its grasp, the faster it starts to choke you, but if you keep still, you can trick the snare into relaxing its grip on you. Thanks to Professor Sprout's herbology class, Hermione is able to escape its clutches and then manages to save a very panicked Ron by conjuring fire to make its tentacles recoil. However, while the fast-moving vines of the Devil's snare are certainly impressive, there are just as many fascinating and deadly plants in real-world botany.

Nepenthes Rajah: King of the Pitcher Plants

First up is something that seems to have climbed right out of the wizarding world. *nepenthes rajah* sounds more like a spell or incantation rather than a plant. Originating in Malaysian Borneo, *nepenthes rajah* is a scrambling vine. This plant comes equipped with giant pitcher-shaped traps, the biggest of which have been known to capture and digest small mammals, frogs, and lizards. It's a slow death, too. When the fated animal falls in, it is drowned and slowly digested by the liters of fluid sitting in the pitcher trap. The trap is essentially a cupped leaf with a waxy, slippery interior, making it difficult to climb out. Scholars have noted that the bodies of small rodents can take months to slowly digest, until finally all that remains within the plant's fluid is the skeleton.

Ominously, the stem of *nepenthes rajah* has distinct climbing ambitions. It usually grows along the ground, but it will try to climb anything it comes into contact with and which can support it. And that stem is formidable. It can grow up to six meters in length, with insects, particularly ants, making up the staple prey of both aerial and terrestrial pitchers.

Dastardly plants have a good side, too. Although *nepenthes rajah* is renowned for trapping unsuspecting creatures, its pitchers are also host to a large number of other organisms. *Nepenthes rajah* plays mutual symbiotic mother to creatures that cannot survive anywhere else, such as the two mosquito taxa named after it: *culex rajah* and *toxorhynchites rajah*.

Bladderworts: Deceptively Innocent

Like devil's snare at St. Mungo's, this plant at first seems small and innocent. But with a name like bladderworts, a botanist might divine there's also something weird afoot. Even though bladderworts is bedecked with pretty flowers, it's actually just as effective at capturing its prey as the *nepenthes rajah* and devil's snare plants. The name bladderworts refers to the bladder-like traps, with which the plant carnivorously captures small organisms. They occur in both fresh water and wet soil, as terrestrial or aquatic species, across every territory on Earth, save Antarctica.

Bladderworts acts swiftly. It can take just ten thousandths of a second for its trap to spring. In the aquatic species, the trapdoor is mechanically triggered, and the prey, along with the surrounding water, is sucked into the bladder. The bladder traps are considered one of the most sophisticated structures in the plant kingdom. But thankfully, its prey is relatively small fry. The aquatic species, common bladderwort, boasts bladders that feed on prey such as water fleas, mosquito larvae, and young tadpoles. By grabbing them by the tail, bladderworts consume tadpoles and larvae by ingesting them, bit by bit.

The Venus Flytrap: The Classic

It's the classic flesh-eating plant of the muggle world. And, with vibrant leaves that close around its prey, the Venus flytrap sets a grasping trap, which is very like the sinister clutches of the magical Devil's snare.

The Venus flytrap is a miracle of nature. People don't normally think of moving plants, but the flytrap can catch insects by its toothed leaves snapping shut, when triggered by prey touching the tiny hairs on the inner surface of the leaf.

The flytrap's mechanism is a sophisticated trap. Imagine a spider, stupid enough to be clambering across the inside of a Venus leaf. If the arachnid triggers one of the tiny hairs on the inner surface, the trap prepares to close. It snaps shut unless it feels a second contact within about twenty seconds of the first strike. This requirement of redundant triggering serves as a safeguard. It means the flytrap doesn't waste energy by trapping objects of no nutritional value. Venus will only begin digestion after five such stimuli, to make sure it's caught a live bug worthy of chomping. The flytrap adds further refinement, too. The speed of snap-shut varies on the amount of humidity, light, size of prey, and general growing conditions. The speed with which the trap closes is a useful indicator of the plant's general well-being, even if the same can't be said for its prey, which includes beetles, spiders and other crawling arthropods.

The Venus flytrap boasts some very famous fans. The founder of geographical botany, John Dalton Hooker, was director of the Royal Botanic Gardens at Kew in London. He shared a keen interest in carnivorous plants with his closest friend, Charles Darwin, who called the flytrap, "one of the most wonderful plants in the world."

The Giant Hogweed: The Late Entry

Giant hogweed is the plant of nightmares. Many plants prove toxic by ingestion, but the *Giant Hogweed*, which grows up to 8 feet tall, can poison you by touch alone. Looking like something from an alien planet, Hogweed poisons with cooperation from an extra-terrestrial body—the sun! As Hogweed is photosensitive, it oozes a thick sap that coats human skin upon contact. At once, the sap reacts with the sun and starts a chemical reaction that burns through flesh. The searing contact can lead to necrosis, and the formation of massive, purple lesions on the skin. Incredibly, the lesions may last for years. Even more worrying is the fact that a minute quantity of sap can cause permanent blindness upon eye contact. It's hardly surprising that giant hogweed plants have become a priority emergency target for muggle toxic plant control departments.

HOW HAS MEDICINE MADE POWERFUL POTIONS FROM PECULIAR PLANTS?

The Anglo-Saxon "watery-elf" disease, thought to be chickenpox, was treated by "mixing herbal lore, magical charms, myth, and religious ideas. The doctor, called the "leech," was a monk and would get the patient to drink a mix of holy water and "English Herbs," while apparently repeating the line, "May the Earth destroy thee with all her might." Like many older cultures, medieval monks believed that the illness was also linked to god and the spirit world.

According to the potions master, Severus Snape, there is a "subtle science and exact art" behind potion making. A potions master specializes in the mixing of various substances to make liquids that can be used to create magical effects in the person that drinks them. Basically, they mess with people's physiology using a mixture of natural ingredients and magic.

According to J.K. Rowling, it wouldn't be possible for a muggle to make a magic potion. Even if they were "given a potions book and the right ingredients, there is always some element of wand work necessary."

Not having a wand hasn't stopped muggles in their quest to produce concoctions to aid in our health or everyday affairs. From noticing the particular effects of eating something, to experimenting with the consequences of combining natural produce in different amounts, this ingenuity has led to powerful potions with extraordinary effects despite the absence of magic. The question is, how have muggles achieved this?

Nature's Medicine Cabinet

Nature holds many chemicals and substances that are harmful to humans, but on the flip side, a great deal of known and unknown natural products can benefit us in our everyday affairs. For example, the South African San people use the hoodia plant to suppress appetite when hunting or doing long journeys. Some other common herbs are ginger, feverfew, evening primrose, milk thistle, ginseng, and St. John's wort.

The reason these herbs are so useful to us comes down to their chemistry. Plants churn out different chemicals to help them function. Primary metabolites such as carbohydrates, vitamins, and proteins are located in all plant cells and are essential to their growth, development, and reproduction. Secondary metabolites, which are derived from primary metabolites but are specific to each plant, are compounds used to attract or defend against other organisms that might pollinate, infect, or try to feed on the plant.

The active ingredients in many of these beneficial compounds of plants can have positive or negative effects on our physiology—effects that druggists often exploit when developing remedies.

There are three types of secondary metabolite that are particularly relevant to medicine: phenolics, terpenoids, and alkaloids, which will be discussed later. Plants produce phenolic compounds to defend themselves against pathogens (disease-causing microorganisms). Salicylic acid, used to reduce acne, is a phenolic compound. A modified version of it is also used to make aspirin. Terpenoids are the biggest group and are primary ingredients in essential oils, which can be toxic to insects while also protecting a plant against bacterial or fungal infection. Some have a potential use against cancer, malaria, inflammation, and various viral and infectious bacterial diseases.

Accumulating Knowledge

The ability to identify the difference between dangerous and beneficial plants is vital to the survival of any life form that eats or interacts with plants. It's not just about recognizing the plant, it's also necessary to identify which parts of the plant are edible or safe to handle. For instance, we can eat an asparagus stem, but the fruit of the plant is poisonous, and while we regularly eat the fruit of a tomato plant, its leaves are poisonous. The more notorious poison ivy and water hemlock plants are poisonous to the touch, as is the manchineel tree, which harbors poisonous fruits and sap.

Without particular technologies to help them, indigenous people relied on sensory cues to determine a plant's potential underlying chemistry. Of course, no method could really provide them with more insight than just trying it and seeing what happens. This knowledge gained through trial and error led to the accumulation of a rich body of knowledge about plants and their effects on the human body. Particular individuals within communities applied this knowledge of plants to create medicines. These herbalists used their skills to prevent, diagnose, improve, or treat physical and mental illness.

The oldest written record of medicinal plant use goes back almost 5,000 years with a Sumerian clay tablet from Nagpur, which references plants like poppy, henbane, and mandrake. There's also the 3,500-year-old Ebers Papyrus from Egypt featuring hundreds of formulas and medicinal remedies that use ingredients such as garlic, myrrh, aloe vera, and mint. A recent study by the Royal Botanic Gardens at Kew in the UK has estimated that there are at least 28,187 plant species in the world currently recorded as being of medicinal use.

Ye Olde Pharmacy

The study of how drugs and medicines affect living systems is called pharmacology, but for millennia it was referred to by the Latin name, materia medica. The term materia medica comes from the title of a book written by the Greek physician Pedanius Dioscorides in the 1st century AD. It was a popular book that was regarded by medical practitioners right through to the Middle Ages.

At that time, the major medicine practitioners in Europe were called apothecaries. These were basically grocers who specialized in herbs, wines and spices. Through the years they became more involved in the storing and selling of confectionery, perfumes, and drugs which they concocted and dispensed to the public. By the mid-16th century they were mainly dealing in substances for professional use by doctors. They were essentially the early equivalent of our modern community pharmacists.

By the 17th century, plant medicine books known as herbals were valued and well-known, such as John Gerard's *The Herball or Generall Historie of Plantes*, and Culpeper's *Complete Herbal*. Culpeper's book was actually used by J.K. Rowling as a source of inspiration for some of the "witchy" sounding plant names used in Harry Potter such as toadflax, flea-wort, mugwort, and knotgrass.

In 1704, it was legally ruled that apothecaries could prescribe and dispense medicines, and in 1815 the Apothecaries Act was passed in an effort to better regulate the field, marking the start of medical regulation. In England, the society of apothecaries were still making and selling medicinal and pharmaceutical products in the 1920s. Nowadays, the apothecary has evolved into the general practitioner, i.e. your local GP.

The rise of chemical analysis in the early 19th century meant that scientists could now extract and modify the active ingredients from plants, rather than just processing whole parts of plant leaves, roots, or flowers. This permitted the creation of specific remedies that wouldn't expose patients to the extraneous compounds that can be present in plants.

From then on, European medical practice became dominated by biomedicine, which applies "the principles of the natural sciences; especially biology and biochemistry." So, how do they get the active ingredients out of plants and into modern drugs?

Drug Development

To obtain the necessary bioactive compounds from the plant, scientists have to extract and isolate them. Starting with a fresh or dried and powdered plant, the required compounds may be soaked (the process of maceration) or filtered slowly (the process of percolation) in water or organic solvents. The solvent affects what compounds can be extracted. For example, tannins and terpenoids can be extracted by water or ethanol whereas anthocyanins can only be extracted by water or methanol solvents. For alkaloids, ethanol or ether is suitable. Whichever process or solvent is used, it's important to ensure that the active ingredients aren't adversely affected, lost, or destroyed.

Extraction can often lead to a variety of compounds being obtained. To isolate these compounds, separation and purification techniques are needed, such as chromatography. Chromatography is a method that allows different chemical properties to move at different speeds while passing through a substance. After a time, particular compounds can be found at different positions along the substance.

Obtaining the active ingredient doesn't necessarily provide the finished drug, though. Active ingredients are often combined with other substances such as sweeteners, preservatives, flavors, lubricants, and vehicles (substances used in liquid or gel mixtures to help carry the active ingredient into the body). These additives, known as excipients, are pharmaceutically inert i.e. they have no particular biological effect on us.

Different ingredients can also be blended together to increase absorption into the body or to target different areas simultaneously. However, it's important to know how these substances decompose over time to make sure that the drug doesn't suddenly become toxic when left on the shelf for a while or if exposed to particular temperatures or substances.

Some drugs may also cause undesired side effects or health and safety concerns, leading scientists to synthesize their own modified versions of natural compounds to provide more suitable forms. For example, chloroform and ether were both adapted to be less liver-toxic or flammable. Other modified drugs include heroin and LSD, which are derivatives of morphine and lysergic acid. These drugs can be ineffective or dangerous if taken in the wrong amounts. A safe dosage lies within a therapeutic window, which is the range of dosages in which a drug will be effective without being toxic. The dosage also has to be adjusted based on the size of the person taking it. The particular therapeutic windows are determined experimentally, and more recently, this has included the use of computer models and tests on particular cells of the body before moving on to animal testing followed by human clinical trials.

Powerful Potions from Peculiar Plants

A pharmaceutical agent of plant origin is called a phytopharmaceutical. An example is salicylic acid, found in willow trees. A modified version of the compound, called acetylsalicylic acid is the active ingredient in aspirin. There's also the aloe-sourced barbaloin, which is an anthraquinone; a class of drug known to have laxative properties. It works by increasing the peristaltic action and reducing water absorption.

By far, the most widely established plant compounds are alkaloids, which are usually bitter tasting and tend to have diverse and powerful physiological effects on humans and other animals. They're also closely linked to psychoactive effects alongside stimulants and hallucinogens. In 1804, morphine became the first alkaloid to be isolated and crystallized.

In the 1950s, rosy periwinkle was found to contain alkaloid compounds that inhibit cancer cell growth. Their discovery and use helped reduce the mortality rate of people with Hodgkin's disease or acute lymphocytic leukemia, which were two of the deadliest cancers at the time. Other well-known plant-sourced alkaloids include the stimulants caffeine and nicotine, cocaine from the Coca plant, whose leaves are a local anaesthetic, and quinine, an anti-malarial compound. Poisonous hemlock and strychnine are also considered alkaloids.

The to-be discovered peculiar plants of the world will likely provide us with more remarkable medicinal compounds with powerful effects on our body. The pharmaceutical industry is huge and although companies put a great deal of money and effort into research and development, ultimately it pays off, warranting the search for more pharmaceutical solutions. One increasingly popular method is bioprospecting where new plant species are sought out for their possible novel compounds. Essentially, that's how peculiar plants have been able to provide medicine with the ingredients to make powerful potions.

THE PSYCHOLOGY OF SEX: DO REAL-LIFE LOVE POTIONS WORK?

Is there a chemistry of seduction? There surely was in the wizarding world. As potions master Severus Snape once said, "You are here to learn the subtle science and exact art of potion-making. As there is little foolish wand-waving here, many of you will hardly believe this is magic. I don't expect you will really understand the beauty of the softly simmering cauldron with its shimmering fumes, the delicate power of liquids that creep through human veins, bewitching the mind, ensnaring the senses. I can teach you how to bottle fame, brew glory, even stopper death."

In the Harry Potter Universe, love potions were brews of infatuation. They rendered the drinker obsessed with the person who gave them the drink. Not only were love potions thought to be very powerful, they were also known to be highly dangerous. Amortentia was the most powerful of such potions, with a mother-of-pearl sheen, and a signature spiral steaming while being brewed.

Like recreational drugs in the muggle world, love potions were banned at Hogwarts. And, like most muggle drug bans, the rules on the use of love potions only seemed to encourage witches and wizards to win hearts by their use. Indeed, even Ron's mum, Molly Weasley, admitted to having brewed a love potion when she was a young witch at Hogwarts. The usual practice was to hide a love potion in food or drink, so the intended victim would know no better.

Getting potions into Hogwarts was like Prohibition. The Weasley shop, Wizard Wheezes, started selling a series of love potions as part of its

WonderWitch promotion. And when Hogwarts caretaker, Argus Filch, banned all their products from the school, Fred and George Weasley instead shipped in potions disguised as perfumes and cough potions. So, the young witches and wizards of Hogwarts traded in contraband, ordering the love potions, despite mandatory searches on owls. Hermione learned evidence of such trade when she overheard girls in the bathroom gossiping about ways to slip Harry a love potion.

Muggles have toyed with the idea of aphrodisiacs for centuries. Foods such as chocolate, avocado, oysters, and honey have all been alleged to carry the reputation of being great for love and fertility. But what about muggle love potions? And what's the chemical possibility that they might actually work?

Muggle Mixtures

The notion of love potions has long been appealing in the muggle world, too. But what progress has been made to synthesize something that might make muggles fall in love? Some scholars believe love potions could soon become a reality. One of the main reasons muggles fall head over heels is that muggle babies simply can't fend for themselves. In contrast, many other animals have offspring quite capable of finding food and being self-sufficient, from the get-go.

Humans have really helpless babies. And that means, from the point of view of evolution, human parents better stay bonded so that their offspring have the best chance of survival. Cue pair-bonding systems. When muggles fall in love, the effect on their brains is unique. Chemical hormones called oxytocin and vasopressin are set free by activating the brain's dopamine system. It's dopamine that builds the bond. Dopamine symptoms are similar to taking a stimulant. Dopamine release fuels the frontal lobe, and causes muggle parents to realize that their partner is someone who should stick around, someone they should feel bonded to.

Muggles miss their partners' smell. And, when separated, their bodies discharge a peptide hormone known as corticotropin-releasing hormone (CPH), involved in the body's stress response. So, scholars think a real love potion could soon exist. Such an elixir could be created from a concoction of oxytocin, vasopressin, and CPH. But chemists don't yet understand exactly how "the delicate power of liquids that creep through human veins" would affect the right part of the muggle brain by stimulating the right systems. In short, a proper love potion won't be available at your local pharmacy just yet, but brain science is developing swiftly. Scholars know far more about the muggle brain than they used to. Not only are they better at understanding the brain, they're also better at modeling muggle brain circuitry. And that means, within a decade, scholars should be in a position to brew the glory of an elixir of love. So, soon, when the love potions are freely available at your local pharmacy, you could take a potion to make yourself fall in love.

However, love potions come with many problems. The archetypal love potion, which makes you desire someone simply because you drank the potion, is certainly fraught with moral dilemma. Ethically, if love potions existed, they'd be subject to the same challenges as so-called date rape drugs, as love potions could be administered to someone without their knowledge. And yet, love potions could also be used to help bond and strengthen long-term relationships. Emotions evolve over time. But a love potion knowingly taken, might be a way of topping up that love that had started to fade.

Would you take the potion?

HAVE SPY AGENCIES USED THEIR OWN VERSION OF A VERITASERUM POTION?

Three things, they say, cannot be long hidden: the sun, the moon, and the truth. The truth is rarely pure and never simple. Yet in the Harry Potter Universe, they still tried bottling it. Veritaserum was a magical and potent truth serum. It fundamentally forced the drinker to answer truthfully to any questions asked of them. Use of Veritaserum was tightly regulated by the Ministry of Magic, who also recognized there were ways of countering the potion, such as quaffing an antidote, or occlumency.

As any proficient potions master would confirm, Veritaserum had a complex chemistry. When carefully synthesized, the serum was not only clear and colorless, but also odorless, which made the potion practically indistinguishable from water. Professor Severus Snape held that the serum needed to mature for a full phase of the moon before use, and stressed that there were other difficulties in its manufacture. The name Veritaserum stemmed from the Latin *veritas*, meaning truth, and the Latin *serum*, meaning liquid or fluid.

The genius of Veritaserum was partly in its chameleon chemistry. Its similarity to water in many of its characteristics meant that it was easily miscible with most drinks. Three drops were a sufficient dosage to make the drinker divulge their innermost secrets. And, in theory at least, the potion worked its magic on the body and mind of the drinker, compelling them to tell the utter truth to any question asked. According, of course, to whatever the drinker perceived to be true.

And yet, the use of Veritaserum had its limits. In courts of magical juris-diction, the use of Veritaserum was considered, "unfair and unreliable to use at a trial," in the same way muggle courts often prohibited the evidence resulting from polygraph tests. As some wizards and witches were skillful enough to repel the effects of the potion while others were not, its use at trial would not necessarily indicate definitive proof of innocence or guilt.

Now, memory is a complicated creature. It's relative to truth, but not exactly the same thing. Wizard kind was aware of the fact that a truth-teller states only what they believe to be true. A teller's sanity and grasp of reality also needed to be factored into discussions. Thus, while the teller's answers may have been heartfelt, they were not necessarily true. Witness the testimony of Barty Crouch, Jr. Some of his replies were true to his mind and memory, and yet his interrogators knew them to be false. Crouch's character was a mitigating factor on the Veritaserum's full effectiveness. Has Veritaserum ever had a parallel in the muggle world, especially in the shadowy realm of espionage?

Truth Serums: Scopolamine

Some kind of super-serum has long been the goal of spies all over the world. Back in the day, brutality was the first weapon of choice. Rather save time and beat seven daylights out of a suspect than do the proper science work, painstakingly looking for evidence in the Sherlockian way. The British were no strangers to brutality, even long before the days of Empire. As Sir James Fitzjames Stephen chronicles in *A History of the Criminal Law in England, Vol. 1*, "They hanged them by the thumbs, or by the head, and hung fires on their feet; they put knotted strings about their heads, and writhed them so that it went to the brain ... Some they put in a chest that was short, and narrow, and shallow, and put sharp stones therein, and pressed the man therein, so that they broke all his limbs ... I neither can nor may tell all the wounds or all the tortures which they inflicted on wretched men in this land."

By the time of the Empire, the British were well practiced. British judge, uncle of Virginia Woolf, and anti-libertarian writer, Sir James Fitzjames Stephen, wrote about torture and truth in *A History of the Criminal Law in England, Vol. 1*, "It is far pleasanter to sit comfortably in the shade rubbing red pepper in some poor devil's eyes, than to go about in the sun hunting up evidence." As chemistry progressed, even rubbing red pepper seemed too much of a chore.

Spooks turned to the "truth serum." A number of drugs have been presumed to relax the recipient's defense so much that they can't help but reveal any hidden truths. Though better than torture, the use of drugs still raises questions about individual rights and liberties, of course. And, as in the magical world, their use has sparked medico-legal controversies over the decades.

The first likely serum was scopolamine, a medication used to treat motion sickness and postoperative nausea. Early in the 20th century, doctors began using scopolamine coupled with morphine and chloroform to create a condition of "twilight sleep" in mothers during childbirth. In 1922, a Dallas obstetrician named Robert House realized that scopolamine might also be used in the interrogation of suspected criminals. In expectant mothers, scopolamine had invoked sedation and drowsiness, a confused disorientation, and amnesia for things that happened during intoxication. Yet the women in twilight sleep also replied to questions not only very accurately, but often with alarming candor! Doctor House came to believe that with scopolamine in their systems, suspects "cannot create a lie . . . and there is no power to think or reason." The idea of a 'truth drug' was launched upon the world.

House published around a dozen papers on scopolamine between 1921 and 1929, and the reputation of House as the "father of truth serum" became so infamous that the very *threat* of scopolamine interrogation was used to get confessions from worried suspects. But the numerous side effects of the drug, which included hallucinations, disturbed perception, headache, rapid heartbeat, and blurred vision, could too easily distract the suspect from the point of the interview.

Truth Serums: Sodium Thiopental

More recently, the sinister truth serum of the movies is sodium thiopental. Although first developed in the 1930s, sodium thiopental is still used today in interrogations by the police and military in some countries. An anesthetic, sodium thiopental is one of a group of drugs known as barbiturates, chemicals used widely in the 1950s and 60s to help better sleep. Barbiturates work by slowing down the speed of messages that travel through your brain and body. The more barbiturates there are, the more the chemical messages slow, finding it hard to jump the gaps between one neuron and another. The speed of thinking slows down very quickly with sodium thiopental. And scientists found that, when sitting in the twilight zone between awake and drugged, the suspect enters a twilight zone and becomes chatty and disinhibited. But, when the drug has worn off, they quite forgot what they had been saying. They could potentially confess and not know they had confessed.

But does sodium thiopental work in interrogation? Research has found that the drug will no doubt make you more inclined to talk. And when under its influence, you are also very suggestible. And that's because the drug is interfering with your higher centers, such as your cortex, where lots of decision making occurs. But there's also a worrying risk you will say whatever your interrogator wants to hear, rather than the truth. The barbiturates may sometimes work in interrogation. But even in ideal conditions they create an outpouring marred by deception, fantasy, and garbled speech. It's still possible, though, for some people to resist drug interrogation, and those likely to withstand ordinary interrogation can hold out in narcosis. No such magic brew as the popular notion of truth serum exists as of yet.

WILL HUMANS EVOLVE LEGILIMENCY AND OCCLUMENCY LIKE SNAPE?

The mind is not a book; not according to Professor Severus Snape, the renowned legilimens. The mind cannot simply be unlocked at will and analyzed with ease. Nor were wizards' thoughts, "etched on the inside of skulls." Rather, the mind is a psychic onion. It is a complex organ of concentric layers. And yet, those who had mastered the necessary powers could still delve into the minds of their prey.

In the Harry Potter Universe, legilimency is the skill of magically passing through the numerous layers of a wizard's mind, and properly fathoming what you found. Wizards, like Snape, who adeptly practiced the art of such psychic probing, are known as a legilimens. To muggles, the skill might be called mind reading. But, naturally, practicing wizards regarded the comparison as guileless.

The opposite of legilimency is occlumency. Wizards use occlumency to shield their mind from the intrusion of a legilimens. Voldemort used legilimency widely, both wandlessly and without words, to enter the minds of wizards. Indeed, Voldemort was regarded as the most accomplished legilimens ever, though that was mainly by his death eaters. Nonetheless, Harry had to be sure to master occlumency to hide his mind from Voldemort.

The very word, occlumency, reminds the reader of the occult. The occult (from the Latin word occultus, meaning clandestine, hidden, or secret) is knowledge of the hidden. Though commonly, occult also refers to knowledge of the paranormal, as opposed to knowledge of the measurable, usually referring to science. So, what does science have to say about such occultist practice?

Fantasy of the Mind

Occultist notions of psychic power have been with us for centuries. Had Isaac Newton not been inspired by the occult concept of action at a distance, he might not have developed his theory of gravity. Newton's use of the occult forces of attraction and repulsion between particles influenced British economist, John Maynard Keynes to suggest that, "Newton was not the first of the age of reason: he was the last of the magicians."

The study of the occult is associated with hidden wisdom. For the occultist, such as Newton, it is the study of truth, a deeper and more profound truth that lies beneath the surface. Much fantasy has been written about this sense of a deeper spiritual reality, which is thought to extend beyond pure reason and the physical sciences. Many writers on this topic believe that re-discovered powers based upon such a hidden reality might be developed in the course of our future evolution.

Psi powers is the name given to the full spectrum of mental powers, which are an assumed element of this hidden reality. The name stems from the study of the pseudoscience of parapsychology, and is a widely used term in the fantasy tradition. In the United States, the term was particularly prominent during the "psi boom" that John W. Campbell Jr. promoted in *Astounding Science Fiction* magazine during the early 1950s. A related term, psionics, derived from combining the psi, signifying parapsychology, with electronics, arose in the late 1940s and early 1950s. Again, Campbell was key. Psionics revolved around the application of electronics to psychical research.

An early instrument used was the Hieronymus machine. Ostensibly, the invention of Dr. Thomas Galen Hieronymus, but promoted widely by Campbell in *Astounding Science Fiction* editorials, Hieronymus machines were mock-ups of real machines. They allegedly worked by analogy or symbolism, and were directed by psi powers. For example, one could create a receiver or similar device of prisms and vacuum tubes by using simple and cheap cardboard or schematic representations. Through the use of psi powers, such a machine would function like its real equivalent. Campbell claimed that such machines actually did perform this way. Unsurprisingly perhaps, the concept was never taken seriously elsewhere. Still, fantasy writers speculated about a future where man could harness such mental capabilities.

Typical is Arthur C. Clarke's *Childhood's End* (1953). The first dawning of a space age suddenly aborts when enormous alien spaceships one day appear above all of the Earth's major cities. The aliens, the overlords, quickly end the arms race and colonialism. Sound familiar? The idea has been copied in movies many times since. In *Childhood's End*, after one hundred years have elapsed in the story, human children start displaying psi powers. They develop telepathy and telekinesis. They become distant from their parents. The overlords' purpose on Earth is finally revealed. They are in service to the overmind, an amorphous extraterrestrial being of pure energy. The overlords are charged with the duty of fostering humanity's transition to a higher plane of existence and merger with the overmind.

Facts of the Mind

Reality check. Interestingly, in the preface of a 1990 reprint and partial re-write of *Childhood's End*, Clarke attempted to unravel pseudoscience from his extraterrestrial message:

"I would be greatly distressed if this book contributed still further to the seduction of the gullible, now cynically exploited by all the media. Bookstores, newsstands, and airwaves are all polluted with mind-rotting bilge about UFOs, psychic powers, astrology, pyramid energies."

So, what exactly is the problem with mind reading? And what are the future prospects of humans evolving a type of telepathy? The nearest thing to telepathy at present on our planet is shark sense. Sharks and several other fish have evolved an electro-sensitivity. They use organs called *ampullae of Lorenzini* to sense spurts of nerve impulses in other fish and worms, as the prey try to bury themselves in seabed sand, away from the predatory shark.

But the reading of thoughts, from brain to brain directly, would need some kind of electromagnetic transmission. And even if there were a potential channel of chatter apart from the electromagnetic transmission, the communicating minds would need to be matching—that is, the identical nerve cell in both brains would have to have the exact same purpose. We already know this is not the case for muggle brains. Not even identical twins, who are commonly conceived to have telepathic powers, have more than most muggles. Even twins have contrasting experiences in their primary years. And these differences program each individual brain with diverse nerve cell links, with numerous types of contrasting connotations. In short, a concept like quidditch will have different nerve nuances from one muggle to another.

The same would go for wizards. Differing brains based on the differing experiences of a lifetime would mean dissimilar mental architectures. Resonance between such minds would make it hard for messages to pass from one wizard to another. Wizard minds would be as different from one another as muggle minds are.

For muggles, technological telepathy might prove more promising. In the future, we might be able to develop a form of psionic wetware—computer technology in which the muggle brain is linked to artificial systems. An internal modem, for example, might make it possible to send messages to another device, planted in another head. This second device would then relay the missive to the recipient. And, to those who knew no better, it might look like telepathy from the outside!

COULD EVOLUTION PRODUCE ITS OWN FLUFFY?

Histor y has many hellhounds. These supernatural dogs from folklore are often guardians to an underworld, the supernatural, or the realm of the dead. And sometimes, even if their name is Fluffy, they guard the Sorcerer's Stone at Hogwarts Castle.

Bought by Hagrid from a "Greek chappie" at the Leaky Cauldron, Fluffy was a huge and vicious three-headed dog whose greatest weakness was to fall fast asleep at the first sound of music. Harry, Ron, and Hermione first met Fluffy in the forbidden area on the third floor of the Castle. As sharp as ever, Hermione noticed that, like other hellhounds of history, Fluffy was guarding something. He stood on a trap door, which the three later deduced to be the way to the Sorcerer's Stone.

When the three students met Fluffy once more, Harry had brought along a flute, so that it could be played to help lull Fluffy asleep. Once the Stone was destroyed, and Fluffy's duties done, Hagrid took Fluffy to the Forbidden Forest, and set him free. Soon after, Dumbledore sent Fluffy back to his native Greece.

The movie version of Fluffy appears to be a breed of English dog, known as a Staffordshire Bull Terrier. And to make the three heads look more realistic, each is given its own personality: one "smart," one "alert," and one "sleepy." But, could evolution produce its own Fluffy?

Many-Headed Monsters

Many-headed monsters have a long pedigree in fantasy. A many-headed Hydra had once born down on a laboring Heracles, son of that "Greek chappie" Zeus. Heracles realized that such a fearsome beast would re-grow any head that was chopped off. Now, a multi-headed creature with re-growable body parts is a wonderful feat of the fantastic imagination. But from where do grotesque ideas like Hydra and Fluffy stem? Could fantasy writers have got the concept from nature itself?

Scholars have recorded cases of many-headed specimens for years. In the 1940s, a two-headed pipefish embryo was dubbed a "tiny monstros-ity." More recently, biologists have seen many examples of two-headed creatures in their labs. Using modern genetics, scholars have come to understand the mutations and cellular displacements that might allow for this phenomenon, and similar ancient cases could have led to the creation of the original myths. Ancient storytellers may have seen such abnormalities, and incorporated them into their tales.

Two-headed, and even three-headed, animals are occasionally found in the wild. The phenomenon, named polycephaly, is not limited to any one class of animals. In recent years, a two-headed bull shark fetus was found in the Gulf of Mexico, and a double-headed dolphin washed up on a Turkish beach. Both are instances of conjoined twins—offspring that develop from an egg that fails to separate after fertilization. Such offspring will often have twin sets of some internal organs and even limbs.

The list of many-headed creatures dates back millions of years. It includes not only snakes, turtles, and kittens, but those ancient monsters that paleontologists have discovered in the fossil record. Evolutionists reckon there are a number of mechanisms that can mean more than one head or face. Heads are an example of convergent evolution. They evolve separately in different groups of species. Heads just seem to be useful adaptations that can arise in a wide range of creatures. And that's why so many sense organs, such as eyes, ears, nose, and mouth, are also found there.

Sonic Hedgehog Gene!

The root of the many-headed phenomenon is at the genetic level. One gene that has a big influence, especially on the width of the face, is pleas-ingly named "Sonic Hedgehog," or SHH. The name is a result of a set of hedgehog (HH) genes, which mutate to make fruit flies born with spiky hair-like structures—that look a little like teeny hedgehogs.

Vertebrates have the Sonic gene. And if the SHH signal is increased during embryonic development, weird things can happen. The head can broaden so much; you get two faces rather than one. We're some way to producing a Fluffy, but not all the way there. The Sonic gene only results in a many-faced, not many-headed specimen. For a totally separate neck and head growing from a single body, a group of cells, known as an organizer, in the early embryo would have to be invoked. Scholars are beginning to understand why such abnormal steps in development happen. And one crucial factor appears to be temperature. For example, one biologist found that higher water temperatures resulted in the development of a two-headed zebrafish embryo.

Nature may well have inspired fantasy, but the reverse is also true—fantasy has inspired science. The ancient myth about Heracles inspired Swedish zoologist, Carl Linnaeus, to name a genus of simple freshwater animals, Hydra. This group of tiny aquatic animals, found by Linnaeus in 1758, are particularly fascinating, as they have many snake-like appendages and even the ability to regenerate like the Hydra in the myth.

Muggles are psychologically prone to be unsettled by natural abnormality. Such reactions help explain why deformity made Fluffy such a frightful creature. Indeed, the description of the two-headed pipefish embryo as a "tiny monstrosity" also serves to show the human discomfort at polycephaly. And that means that Fluffy is far from being the only many-headed monster in human mythology. There is an ancient Japanese myth, yamata no orochi, which tells of an eight-headed snake. A Slavic myth, zmey gorynych, speaks of a three-headed dragon. And, of course, another beast Heracles had to labor against was the many-headed dog Cerberus.

So, in many ways, evolution has already made its own Fluffy. Science still has much to discover about polycephaly in our planet's many creatures. But given the low survival rate of many-headed organisms, both in the wild and in captivity, the trait is likely to remain an extraordinary and unsettling sight for muggles everywhere. Such creatures represent many challenges in one, an adversary that cannot be easily defeated. And the whole creature culture of such multi-headed monsters is likely to have a long life, unlike polycephalic creatures in the wild.

WHERE AND WHEN MIGHT HARRY FIND DRAGONS?

The smooth scarlet scales of the Chinese Fireball. The yellow eyes and bronze horn of the Hungarian Horntail. The black spine of the Norwegian Ridgeback. And the copper-colored scales of the Peruvian Vipertooth. Dragons played an imaginative part in the Harry Potter Universe.

Even the motto of Hogwarts School was *draco dormiens numquam titillandus*, or, never tickle a sleeping dragon. The Hogwarts gamekeeper, Rubeus Hagrid, truly adored dragons. For a short time, Hagrid cared for a Norwegian Ridgeback named Norbert. When Norbert turned out to be female, (s)he was swiftly renamed Norberta.

In the wizarding world, various valuable resources were drawn from dragons. The challenge came with actually getting at those resources, as it needed over a dozen wizards just to stun a dragon. For fear of being seen by muggles, who believed them to be mere myth, dragons were kept on special reserves about the globe, far from human habitation. Dragons could not be domesticated, despite individuals trying to do so. Wizard zoologists who specialized in dragons were known as dragonologists.

As folklorists will confirm, dragons in fantasy often possess traits and characteristics of many other creatures. Those from India might have the head of an elephant. Those from the Middle East might have the traits of a lion, or bird of prey, or the numerous heads of serpents. And the body color of dragons, ranging from green, red, or black to the rarer yellow, blue or white, echoes the habitat of the culture in which the dragon is imagined. But, if Harry was looking for a dragon, where and when might he find one?

A History of Dragons

Dragons are maybe the most enduring creatures of fantasy. They adorn the flags of Wales, Bhutan, and Malta. They were also featured on the Chinese flag during the days of the Qing dynasty. They're known in many global cultures, today populating film, Tolkien-esque fiction, and video games. But their history is a long and ancient one.

Little is known of the time when and where the stories of dragons first emerged. But by the time of the ancient Greeks and Sumerians, tales were already being told of huge and draconian flying serpents. Time was when history viewed the dragon with a modicum of balance. Like other fantastic beasts, they were often genial and protecting, but, like many wild animals, they could at times prove dodgy and dangerous. Tales of kindly dragons seemed to vanish in a puff of smoke with the global spread of Christianity, when dragons took on a decidedly devilish air and their sinister ways came to stand for Satan.

By medieval times, most folk got their ideas about dragons from the Bible. Indeed, most pious people believed in the literal existence of dragons. Witness the evidence of the draconian leviathan from the Book of Job, chapter 41:

"I will not fail to speak of Leviathan's limbs, its strength and its graceful form. Who can strip off its outer coat? Who can penetrate its double coat of armor? Who dares open the doors of its mouth, ringed about with fearsome teeth? Its back has rows of shields tightly sealed together; each is so close to the next that no air can pass between. They are joined fast to one another; they cling together and cannot be parted. Its snorting throws out flashes of light; its eyes are like the rays of dawn. Flames stream from its mouth; sparks of fire shoot out. Smoke pours from its nostrils as from a boiling pot over burning reeds. Its breath sets coals ablaze, and flames dart from its mouth."

Dragons became one of the few creatures of fantasy that were cast as potent and powerful, and a worthy and awesome enemy to be slain. The Christian church created myths of righteous adventurers and sincere saints, on quests to seek and vanquish dragons, a fitting symbol for Satan. Dragons became synonymous with the breath of fire.

Medieval artists, such as Dutch genius Hieronymus Bosch, depicted fire-breathing dragons over the mouth of hell. Look closely at the right panel of Bosch's *Garden of Earthly Delights*, painted in the early 1500s, and you might spot the odd dragon, flying high above the pits of hellfire. The Gates of Hell were often depicted as the mouth of a monster, the smoke and flames of Hades belching out. For the devout who believed in the literal reality of hell, the existence of Satanic dragons wasn't such a stretch.

After all, this was a time when people believed in witches and were-wolves, angels and demons, and heretics and persecution. In 1458, a pig was actually hanged for murder in Burgundy. The French judge Henri Boguet said in 1602 that an apple was possessed by demons. And a few years later, Italian Jesuits attempted to calculate the physical dimensions of hell. Strange times, indeed.

Here Be Dragons

Is there evidence of a link between dragons and real creatures? Possibly. The belief in dragons wasn't simply conjured out of thin air. There was hard evidence in the form of giant bones, which were on occasion unearthed in various parts of the globe. For millennia, few people knew what to make of them. Over time, dragons became the guess of choice for those with no knowledge of dinosaurs.

The word dragon derives from the ancient Greek *draconta*, which means "to watch." Here is the origin of the notion that dragons guard mountains of gold—or Gringotts. No one seemed to wonder why a mythical creature as powerful as a dragon might need coins to pay for anything. Perhaps it was merely the reward received by those brave adventurers who managed to vanquish the mighty beast.

Today, few would believe that so huge and fantastic a fire-breathing creature as a dragon may lurk in some lost land, awaiting discovery while inhabiting some uncharted skies still unseen. But, just a few centuries back, it was believed that dragons were finally discovered. Sailors returning from Indonesia told tales of the Komodo dragon. Destructive, deadly, and reaching ten feet in length, might the Komodo dragon be a flightless cousin of more exotic beasts elsewhere? The myth was helped by the belief that the bite of the Komodo dragon was deadly. Its very breath was toxic. The myth stood until 2013, when a team of scholars from the University of Queensland discovered that the Komodo mouth was no fuller of toxic bacteria than those of other carnivores. The dragon is clearly a chameleon of the imagination.

Scholarly research into a natural history of dragons suggests that a huge cornucopia of creatures influenced the modern idea of a dragon. Traits and characteristics of huge snakes and hydras, gargoyles and dragon-gods, as well as more obscure beasts, such as basilisks, wyverns and cockatrices, worked their way into what we now think a dragon might have looked like. Though most folk might easily imagine a dragon, their notions and descriptions of dragons vary dramatically. Some might picture winged dragons; others plant them squarely on terra firma. Some dragons are given voice, or breathe fire; others make them mute and smokeless. Some might be measured in mere meters; others span a measure in miles. And some dragons are pictured in a submarine world, while others are found only in caves on the highest hills. So, if Harry were looking for fabulous creatures, draconian in sheer size and variety, his best bet would be to travel back to the days of the dinosaur, when they became the dominant terrestrial vertebrates, over 200 million years ago.

WHY WOULD PIGEONS, AND NOT OWLS, MAKE A WIZARD'S BEST FRIEND?

They ghost through the air on shadowy, moonlit nights. They make their muffled way through the dark skies on their way to and from the Owlery at Hogwarts Castle. And they glide on nameless winds, their melancholy cries the only hint that they carry messages for wizard kind.

Owls are as abundant in nature as they are in the magic world of the *Harry Potter* series. Owls are found across the planet, with over 200 species of this mostly solitary bird. Nocturnal birds of prey, owls are typically upright in posture, with binocular vision, and feathers evolved for silent flight. Rarely do they appear in the day.

The Eurasian eagle-owl, such as that sported by the Malfoys, has a wingspan of nearly two meters, and can eat foxes, herons, and even small dogs. So at least some owls are large and strong enough to carry parcels. Though they are certainly clever enough, owls have never been used to deliver messages.

Humans have held owls in high regard since before the beginning of civilization. The Chauvet-Pont-d'Arc Cave in the Ardèche department of southern France contains, among its wonderful display of figurative paintings, the clear image of an owl, engraved into the rock. The owl is shown on the cave wall with its head seen from the front, but its body seen from the back. Little wonder that even pre-historic cultures associated owls with supernatural powers. Early humans may also have been fascinated by the owl's other inhumane ability to see in the dark, such as the cave itself. The Chauvet cave owl painting is thought by scholars to be at least 30,000 years old. It's a long pedigree in magic.

Though owls are as yet unused by muggles as messengers, nature provides some stunning alternatives.

Arctic Terns

The bird with the most spectacular stamina is the Arctic tern. This seabird is mightily migratory. It sees not one, but two summers each year, as it migrates along a route from the Arctic north to the Antarctic coast for the southern summer, only to follow the same path once more, six months later. Recent muggle studies suggest an annual round-trip of around 44,000 miles for terns nesting in Iceland, and an incredible 56,000 miles for terns nesting in the Netherlands.

Arctic terns are by far the longest migratory creatures in the entire animal kingdom. Not only that, but the Arctic tern is a long-lived bird. Many reach fifteen to thirty years of age, in some cases outliving many wild owls. The species is also abundant, with an estimated population of one million birds across the globe.

But the Arctic tern is also unproven as a message carrier. The most distinguished pedigree as nature's dispatch rider must surely go to the common columbine.

Using the Columbine

Consider the candidacy of the humble pigeon. Pigeons, along with doves, belong to the bird family, *Columbidae*. The family includes over 300 species, and is probably the most common bird in the world. The species to which we refer most as 'pigeon' is the rock dove. Pigeon flights as long as 1,000 miles have been recorded, and their average flying speed is about the speed of a car. Pigeons have a long history with humans as carriers. The Egyptians and the Persians first used them to carry messages, around 3,000 years ago, the Republic of Genoa set up a system of pigeon watchtowers that ran along the Mediterranean Sea.

But how do pigeons know where to fly? Nobody knows for sure. Every seemingly reasonable hypothesis has been tested to destruction. Some muggle studies suggested 'magneto-perception'—that pigeons have a kind of map and compass in their heads. This perception means they sometimes use the sun to work out where they're going and, since the Earth is like a big magnet, they may also use the Earth's magnetic field to get them home. When they get close to where they're going, the theory supposes, they also use landmarks.

Believing that the pigeon perception was magnetic, muggles used pigeons as messengers in World War II. Pigeons were routinely taken on Lancaster bombers, and together the birds contributed to the war effort as the British Royal Air Force Pigeon Corps. The idea was that if the pigeons were ditched in the North Sea, on the way back from an aerial sortie over Germany, the flight navigator would tie a map reference of where the plane last was to the leg of the pigeon, in case radio contact was broken. These pigeons saved thousands of lives.

War Heroes

But were the pigeons using magnetism and landmarks? Some were released in the middle of the night in freezing fog, 100 miles from land, with no landmarks in sight, and they still got home. The most outstanding of the pigeons were awarded medals by the British. The meritorious performance list has about 500 examples of such astonishing feats. They were literally dropped out of planes in the middle of nowhere, often during cold winter nights, but still got home the next morning.

To get to the bottom of pigeon perception, muggles went to work. They blocked the pigeon's nostrils up with wax. Turpentine was placed on their beaks to disorient their sense of smell because they figured the mysterious sense of the birds might be olfactory. They even severed the olfactory nerves of the pigeons, in some dubious experiments.

But the muggles were undeterred by their lack of success. They strapped magnets to the birds' wings. They even wound Helmholtz magnetic coils about the pigeons' heads. Even when the muggles fitted the birds' vision with frosted glass contact lenses, and released them over 200 miles away from home, they still flocked down within a quarter of a mile of the loft!

Even if the pigeons are released on cloudy days, or if their internal clocks have been shifted 6 hours, or 12 hours, by keeping them under artificial day-lengths for weeks, they still come home.

Every one of these muggle experiments aimed to test whether the pigeons work via an internal compass or by recognizing and using landmarks on the ground below. One muggle experiment even anaesthetized the pigeons and placed them in rotating drums in an effort to throw their sense of direction. Still, on release, they flew straight home.

Home Ties

For over one hundred years, pigeons have remained something of a mystery. Charles Darwin had suggested in 1873 in a paper on the origin of instincts in *Nature*, that pigeons might commit to memory their journey out, and somehow repeat it on the way home. All evidence to date suggests the means by which the pigeon navigates remains a mystery. Yet, there seems to be some unknown form of connection between the pigeon and its home.

As almost all previous experiments involved moving the pigeons from their home, a new set of experiments moved their home instead using a mobile pigeon loft. Once more, there's a history here, as the British Pigeon Corps employed mobile lofts behind the front lines in World War I. Like the Night Bus in the magical world, the mobile pigeon lofts were fashioned out of London buses, converted especially for the purpose.

When the lofts were first moved, just a half a mile away, the pigeons seemed totally confused. Though they could see the slightly moved loft, they encircled the area, flying about the place where the loft *used* to be, for several hours. Just as anyone would be confused, if they'd found their home had been moved 100 yards down the street, while they were away. Eventually, the bravest of the birds would try out the new loft location by simply diving in. After the loft had been moved several times, the rest of the pigeons returned home.

The secret of the pigeon's long and prestigious pedigree remains a mystery. It would seem the perfect bird for the wizarding world. Making their way with the latest wizard word, come hell or high-water, from home to Diagon Alley, and to a huge Hogwarts Loft in the Castle's West Tower.

IS IT POSSIBLE TO STUPEFY SOMEONE?

It's 2011, the location is London, England, and you're a police officer taking a well-deserved break to scoff your lovingly prepared sandwiches. Your shift will be over in a couple of hours and you're desperate to get back to your family. Then there's a call on your police radio.

"Youths are looting shops and burning cars in Tottenham. All available units required."

You drop your sandwich and respond, immediately heading over to the scene with your colleague. As the situation literally heats up, you find yourself surrounded by rioters, most of whom appear younger than your own kids. Fearful for your safety you draw your weapon and get ready to engage. Situations like this are real for the thousands of law enforcers across the globe, and with the prevalence of terrorism there has been a call for greater security.

In such a profession, it's clear to see that there is a real need for safe methods to incapacitate these would-be aggressors, without causing them serious harm. Especially when they are not legal adults and not legally responsible for their actions. Clearly a muggle equivalent of the *Stupefy* stunning spell is in order. But how possible is that?

Stupefy!

On J.K. Rowling's *Pottermore* website, the stupefying charm is described as a useful spell used to knock out an opponent in a duel by stunning them or rendering them unconscious. You don't need to be a wizard or cast a spell to create that effect in the real world though.

The stupefying charm is like the wizarding equivalent of a boxer's knock-out punch. In boxing a knock-out is simply a blow that incapacitates an opponent, whether immediately with a loss of consciousness, or after receiving a strong body blow that prevents an opponent from continuing.

Although a knock-out can be caused by a blow to any part of the body, a shot to the head is the one that usually springs to mind. When struck, the blow can cause the victim to lose consciousness, falling to the ground in a slump. However, if the casting of the stupefying charm worked in this way, the wizard on the receiving end would be hard pressed to hide the physical after effects.

What exactly is happening inside a person's head to actually cause a knock-out?

Head Trauma

The brain is a fragile organ with more than 100 billion nerves connected via trillions of synapses. Thankfully, the brain is surrounded by the cranium which forms a hard shell to help protect the brain from things like contamination, penetration, and deformation.

However, while the skull may be able to take an impact, the brain can still get rattled around inside as a result of a sudden, forceful movement. This is because there is a fluid filled space between the brain and skull. Depending on the force of the impact, the brain can receive trauma in the form of concussion, where a person may feel dazed or even experience a loss of consciousness for a few seconds or minutes.

Concussion is the most common type of brain injury. As the brain wobbles, contorts, and ricochets against the inside of the skull, stretching the blood vessels, damaging the cranial nerves, and killing brain cells, multiple neurotransmitters in the brain also fire off simultaneously, overloading the nervous system into a temporary state of paralysis. As the muscles subsequently relax, the person collapses to the floor.

It's also possible for an impact to disrupt the flow of blood and oxygen to the head, again leading to a blackout. In any case, you can expect the victim to suffer negative effects such as headaches, confusion, mood changes, and memory loss. Rest is considered as the best thing for recovery but even then, it can take a few months or even years for a concussion to heal fully.

Clearly, stupefying through concussion would leave most wizards with heavy heads. What about a less traumatic means of knock out?

Knock-Out Substances

For this, we need look no further than a medical operating room. In surgery, it's important that a patient stay completely still and relaxed while the surgeon makes their precise incisions. A cocktail of drugs is used to achieve the desired state in the patient. These are administered by an anesthesiologist who typically provides sedatives to make you sleep and analgesics to help with the pain. Generally, anesthetics are used to reduce sensation, which includes pain, whereas analgesics only deal with the pain.

As *Stupefy* acts to effectively render a person unconscious, it's more like a sedative than a straight up anesthetic. In surgery, they use general anesthesia to induce unconsciousness, although it can also be used to remove sensation in a conscious patient. It can be administered directly into the veins, or inhaled as a gas through a breathing mask.

Under general anesthetic the patient often requires a breathing tube to be inserted (intubation) into the trachea to assist in breathing and protect the lungs. Side effects following general anesthesia can include a low chance of vomiting and/or nausea.

Some substances that have historically been used to render people unconscious include *Trichloromethane* (AKA chloroform) and *Diethyl Ether*, often just referred to as ether. These methods of delivery all require direct, close-up administration by somebody. The implications are that a muggle would have to get up close to administer their *Stupefy* charm equivalent, whereas wizards can stun from a distance.

Do we have anything that could provide the same advantage?

From a Distance

Considering the aforementioned ways in which the brain can be rendered unconscious i.e. through head trauma or chemically, there are a few ways that a person could be knocked out from a distance. Some that are more immediate and others that would take a little longer to produce their effect.

Regarding head trauma, it's possible that this could result from a so-called non-lethal projectile such as a bean bag or a rubber, plastic, or wooden bullet. These are fired from regular or specially modified guns. However, in real situations the use of these projectiles have led to some fatalities when fired at close range or towards a particularly vulnerable part of the body.

Less dangerous may be to fire a tranquilizer dart instead. These are used frequently for catching animals, where the dart is basically a drug-filled syringe that empties into the animal upon impact. The injection isn't into the veins either, it's actually absorbed through the muscles, which means it can take anywhere from a few minutes to more than half an hour for the animal to feel its effects.

This isn't a realistic option for humans though. The dose must be strong enough to be effective but not so strong that it overdoses the target. Also, the clothing the target wears can affect the way the dart impacts their body. So, whenever we see tranquilizer darts used in movies, the immediate effects are more artistic license than reality.

So how about another staple of the storytelling industry, namely throwing in some sleeping or knockout gas? Well, it would first need to be potent enough to affect any individuals within a certain proximity. Additionally, it would have to be released in a way that immerses relevant targets in enough of the substance to cause the necessary stupefying effect, but this has proven to have some problems.

In 2002, Russian special forces stormed a Moscow theater to free 850 hostages from 40 or more captors. Before the special forces entered they pumped a sleeping gas into the theater with hopes to subdue the occupants. Unfortunately, 130 hostages lost their lives due to the effects of the gas. It's thought that they suffocated as their unconscious state meant that they couldn't adopt a necessary posture to breath properly.

Is It Possible to Stupefy Someone?

The answer is a resounding yes. Although it'll probably involve a head trauma or a drug-filled infusion. However, in the case of a head trauma there's no guarantee that you won't cause the person brain damage or other problems as a result of it. In the case of medical substances, they would have to be administered under carefully controlled conditions to avoid any adverse effects. So, although it's possible, there's little likelihood that a police officer could get away with deliberately choosing this method to incapacitate an aggressor.

And just in case you were thinking that a taser (AKA stun gun) was the obvious answer to this question, it's not. A taser works by disrupting the electrical activity within the muscles between the taser probes. Once it has disconnected, the target can move again, unless they have suffered any unforeseen side effects. While it stuns, as per a stunning charm, it does not render a person unconscious. Unless, of course, the person was unfortunate enough to fall and suffer head trauma.

WHY ARE DEATH EATER "PURE-BLOODS" WRONG ABOUT BREEDING AND THE GENE POOL?

S nobbery harbors seeds of both hypocrisy and despair. In the Harry Potter Universe, snobbery drove the desire of pure-blood wizard families, who regarded themselves as superior to those witches and wizards with muggles in their family trees.

The very name pure-blood refers to a wizard family with no muggle or non-magic blood. The origin of the idea was synonymous with Salazar Slytherin, one of the four founders of Hogwarts School of Witchcraft and Wizardry. Slytherin's loathing to teach any muggle-born student led to a rift with his three fellow founders, and his eventual departure from his job and the school.

Pure-bloods were rarely, if ever, what they seemed. Claiming purely magical heritage, in reality, so-called pure-blood families argued their alleged purity by denying or lying about the muggle-borns within their family trees. Preferring to keep their bloodline pure, the families bred only with other pure-bloods, and not the muggle-born "mud-bloods," a very derogatory term in the wizarding world.

In denial and despair at a changing world, pure-bloods then tried to heap their hypocrisy upon wizard kind by suggesting that any witch or wizard who mixed with "dirty blood," was a "blood traitor." In truth, there was not a witch or wizard whose blood had not at some time mingled with that of muggles. Indeed, if there were no muggles in wizarding family trees, the wizarding race would have died out long ago, as the number of pure-blood families was declining, and their blood type the least common in the magical world.

The house of Black was a case in point. They were a typical pure-blood family who claimed they could trace their pure-blood status through many generations of magical ancestors. They denied their tree contained any muggles, and their family motto was *Toujours pur*, or "Always (or Still) Pure." But what does science have to say about the views of such pure-blood wizards and so many of the death eaters, whose leader denied the existence of his "filthy muggle father?"

Self-Help for Humans

The muggle world is full of self-help books, offering advice on how to improve oneself. Maybe they think you need more exercise. Or perhaps you need to read more books. Maybe it's just a case of cutting down on visits to the local burger bar. But the topic of eugenics goes one better. Eugenics is about improving the human race as a whole, through blood-lines. Odd word, eugenics. The *eu* bit is Greek and means good; the *gen* bit refers to birth or race. Together, they make a word that suggests the quality of the human population can be improved.

It sounds fine, in principal. But the real trouble is that eugenics has a history of very dodgy ideas. One of the basic principles of eugenics is opposition to miscegenation or the mixing of races. Death eaters believe they are superior to all others, and argue against miscegenation, using the insult mud-bloods for those with mixed blood, which includes Hermione.

Eugenics has a horrible history. When people realized humans inherited traits from their ancestors, they began to come up with ideas to "improve" the human race. Take Greek philosopher Plato. In his book, *The Republic*, Plato argued that a good way to improve the human race was to kill inferior babies at birth. Hardly a vote winner. Plato was opposed by another Greek philosopher named Hippocrates (the founder of medicine), whose Hippocratic Oath is still taken by doctors today.

The first tale to explore a society based on eugenics was *Gulliver's Travels*. This early fantasy story was written way back in 1726 by Irish writer

Jonathan Swift. At one point in the story, Gulliver arrives in the land of the Houyhnhnms. These creatures, which are identical to horses, run a eugenics program involving the selective breeding of their human slaves, known as yahoos. At first, Gulliver is mistaken for one of the yahoos, but he manages to convince the horse masters that he is intelligent enough to be saved. If he hadn't spoken up, Gulliver would have been sacrificed as part of their cruel eugenics program. When *Gulliver's Travels* was written, improving human bloodlines wasn't known as eugenics.

It was the cousin of Charles Darwin, Francis Galton, who coined the term eugenics. Some of Galton's earliest research was based on obituary entries in *The Times* of London. By studying the entries, Galton claimed to trace what he saw as superior human qualities being passed down from generation to generation among Europe's most eminent men. In contrast, he suggested that weak, inferior, and even dangerous traits were also being passed down—most clearly, in Galton's eyes, in society's lower classes, and within certain races.

Galton believed in the inequality of humans. For example, he thought Africans were inferior and suggested that the East coast of Africa be settled by the Chinese who were, according to Galton, superior. Galton's theories were set out in his book *Hereditary Genius*, published in 1869. His plan was twofold. Galton's positive eugenics proposed a human breeding program to produce superior people, and his negative eugenics urged the improvement of the quality of the human race by eliminating or excluding biologically inferior people from the breeding population.

It's a small step indeed from the science of Galton to the chilling eugenics of the Nazis. During the 1930s and '40s, the Nazis forced hundreds of thousands of men, women, and children to be sterilized to prevent them from passing on their genes. Much of the horror of the Nazis regime was foreseen in a 1937 fantasy book called *Swastika Night* by British writer Katherine Burdekin. The novel bears striking similarities to George Orwell's *1984*, published more than a decade later. In both books, the past has been repackaged, and history rewritten. Language is distorted, few books exist apart from propaganda, and a secret book is sole witness to the past. In her future history, Burdekin feared the Nazis would come to dominate the world, and force their ideas of inequality on the human race. *Swastika Night* almost came true. But thankfully for us all, the world pulled together and the Nazis were defeated, along with their breeding programs and genocide.

Breed Outside your Local Pool

It makes total sense for wizards and muggles to have interbred. Making families within a select circle can be downright dangerous. Most people have a few hidden genes that can cause deadly diseases, but it's not normally a problem because we carry two copies of each gene, one from each parent. So, as long as one of the two gene copies is fine, you don't usually get the disease. Bad genes kick in only if both parents pass it down to their offspring. And that would have happened more often when wizards were more closely related.

Indeed, some wizards might be dicing with danger. The more closely related a wizard and witch are, the more probable it is they shared the same bad genes. In such cases, each child can have up to a 25 percent chance of contracting the disease. And it's for this reason that inbreeding is outlawed in many muggle countries.

Consider two more examples, one wizard and one muggle. Pure-blood wizarding families, such as the Blacks and Gaunts, practiced marriage between cousins to keep their pure-blood status. They disowned any family members who married a mud-blood, and yet the Gaunts suffered from problems of inbreeding. As a result, family members showed signs of violent tendencies, mental instability, and even diminished magical capability. In the muggle world, old royal families of European inbreeding were also a problem. The Habsburgs ruled large parts of Europe for many centuries. But after many marriages between first cousins and even uncles and nieces, when King Charles II was born and inherited "bad genes," meaning he had physical and mental disabilities and likely could not have children, the Habsburg rule was ended.

In 2015, one of the largest studies to date into genetic diversity went one step further. The study looked at the genetic background and health of more than 350,000 people, from about 100 communities across four continents of the muggle world. It found that the children of parents who are more distantly related tend to be taller and smarter than their peers. They also found that height and intelligence might be increasing, as a growing number of people are marrying people from more distant parts of the world, which may explain the increase in intelligence from one generation to the next, as documented in the 20th century. Pure-blood wizards had it all wrong. Mud-bloods will inherit the Earth!

PART IV
MAGICAL MISCELLANY

PLATFORM 9¾: ARE THERE REAL HIDDEN RAILWAY STATIONS IN LONDON?

The mighty scarlet engine awaits, as white clouds of steam billow from the chimney—a stepping-stone to a magical far-off destination. All windows down, all pistons poised, all sense of being in a hurry gone, the Hogwarts Express will soon pull out. A fleeting presence in a momentary neighborhood, and soon the Express will run where the lake's level breadth begins, where sky and water meet. Soon.

For now, though, the scarlet engine sits on a platform that denies its own existence: Platform 9¾, King's Cross Station, London, England. Magically hidden behind the barrier between muggle platforms 9 and 10, platform 9¾ is where students board the train for Hogwarts School of Witchcraft and Wizardry. An inquisitive young wizard might wonder what other fractional platforms lay between their whole-numbered counterparts. Why stop at 9¾? Perhaps there's a platform where a magical version of the Orient Express waits to whisk passengers off to wizard-only villages in continental Europe. Or maybe on another there's a quadrennial special to the Quidditch World Cup.

Platform 9¾ was the brainchild of then Minister for Magic, Evangeline Orpington, in the 1850s. The Ministry of Magic had long pondered the age-old problem of how to convey hundreds of students to and from Hogwarts every year, without attracting attention. They had procured the *Hogwarts Express* in the mid-1800s, surely a magnificent engine for its time. And a railway station had been built at Hogsmeade, in excited expectation of the engine's imminent arrival.

But there remained the challenge of building a railway station in the middle of London. Surely that would strain even the muggles' infamous resolve to miss the magic, which was exploding right in front of their very faces. So, a solution was struck: a hidden magical platform would be concealed within the brand new, muggle-built King's Cross Station, reachable by witches and wizards only.

This cunning conceit, to secret a railway station in plain sight of a busy metropolis, makes one wonder what other stations might be concealed in the labyrinth that is old London town.

Railway Revolution

In a curious kind of magical way, it was photosynthesis that had led to Britain's steam-powered Industrial Revolution. Millions of years before, during the clammy Carboniferous, plants on Earth had absorbed the sun's energy, taking carbon from the carbon dioxide in the air, and using it to create living tissue. When the great plants of the Carboniferous died down into the Earth, their energy became frozen in time. Coal preserved sunlight, and when the British began to burn coal, its fire was the many years of sun-fire now set free from the tree. Coal was, and remains, the frozen sunshine of buried forests.

It was the thirst for fire-power that had led to the steam engine. The engines were first designed to pump water out of mines, to get at the fiery coal. Steam engines quickly became the first kind of engine to be widely used, and the spirit of their modern day. It powered all early locomotive trains, steamships, and factories. And over the next two centuries, steam power would forever change the world.

Those who built the trains and railroads were the shock troops of industrialization. Railways opened up countries and continents to capital. The rapidly expanding rail-lines spread from Britain like the giant web of a mechanized beast. And, at the very center of this huge machine, the heart of the circulating veins and capillaries of commerce, sat Victorian London. During the 19th century, London grew greatly. But the burgeoning development of a commuting population led to traffic congestion problems. Horse manure became a major concern.

Victorian London typically had 11,000 cabs and several thousand omnibuses. Each mode of transport used several horses, so the city had more than 50,000 horses in public transport alone, with each animal producing 15–35 pounds of manure a day. One commentator remarked, "How much pleasanter the streets of a great city would be if the horse was an extinct animal,".

Sweepers were used to clear paths through the dung, which was usually sludge in the wet weather of London, or a fine blown powder on the odd dry day. But the piles of manure attracted huge numbers of flies. One estimate suggests that three billion flies hatched in horse manure per day in such cities, with tens of thousands of deaths each year blamed on the manure.

It gets worse.

The horses produced tens of thousands of gallons of urine daily, were amazingly noisy (their iron shoes on cobbles made conversation intolerable on the bustling streets), and were far more dangerous than modern motorized traffic, with a fatality rate 75 percent higher per capita than today.

The problems didn't disappear when the horses died. The average working horse had a life expectancy of only three years. Scores died each day and, as dead horses were hard to shift, street cleaners would wait days for the corpses to rot, so they easily be sawed into chunks.

Going Underground

In short, London was a city desperately looking for a traffic solution! The train was hailed as an environmental savior. By the middle of the century, there were seven railway termini located around the urban center of the metropolis. And soon the idea arose of an underground railway, linking the City of London with these satellite stations.

Visitors to London will be familiar with its now long-established Underground. The tube was the world's first subterranean railway. Today, over one hundred miles of underground, networked track serve around four million passengers a day, one of the largest on the planet. But, once in a while, 'ghost' stations are unearthed.

Engineers recently uncovered the remains of a lost station that shut a century ago in south London. Long-forgotten Southwark Park station was open for a little over a dozen years, from 1902 until its catacombs were closed for good in March 1915. Southwark Park was one of several stations in the metropolis that closed down due to the growing popularity of trams and buses, and the outbreak of World War I. It ferried commuters between the London Bridge and Greenwich.

With dystopian shades of the lugubrious steam-punk city that sits above, and a heavily tiled original ticket hall, which sits in the arches of a viaduct, Southwark Park's creepy corridors and eerie atmosphere are reminiscent of the *BioShock* video game series. Urban explorers have snapped haunting photographs of other abandoned and long-forgotten London Underground stations. Lying deep below the city are disused platforms and derelict stations, which snake for miles underground.

Among them is the dusty Aldwych Underground station, it was closed in 1994 but has since been used as a set for several high-end movie and television productions, including *Sherlock*, *Mr. Selfridge*, and *V for Vendetta*. But perhaps the most enduring London Underground legend is the story of secret government tunnels, used during World War II.

During the war, the number of operable telecoms exchanges in London was very limited. One of the main exchanges was located in the City of London, a fair distance from Whitehall, which housed the War Office, the department of British Government that ran the British Army between the 17th century and 1964, when its duties were handed over to the Ministry of Defence.

As phone lines above ground proved impractical, a hybrid network of tunnels was created under Whitehall, using tube-tunneling techniques. Although this secret government tunnel was actually just a service tunnel, it also served as an 'escape tunnel', a potential route between Whitehall buildings in emergencies, such as gas attacks. Much of the detail of this secret tunnel network under Whitehall is locked away in the National Archives awaiting declassification.

So, put a note in your diaries for 2026. That's when the fact behind further rumors will be released, as the Whitehall tunnels were reportedly also linked to the deep level telecoms tunnels constructed during the Cold War in case of imminent attack.

HOW COULD YOU MAKE A ROOM OF REQUIREMENT?

S ituated on the seventh floor of Hogwarts Castle is the most remarkable of rooms. Boasting a secret entrance opposite a tapestry of Barnabas the Barmy trying to teach trolls the basics of ballet, the Room of Requirement was seemingly like Schrödinger's Cat, both there and gone at the same time. A select few visitors to that corridor knew that the way to unlock the Room was to walk past it three times, thinking about what it was you needed. Only then would the door appear.

The Room of Requirement, which was also known as the "Come and Go Room", was believed to have a certain degree of sentience. This belief was based on the evidence that the room converted itself into whatever the witch or wizard needed it to be at that moment in time. Though, understandably, there was a list of limitations.

Interestingly, it was also thought that the Room was unplottable. It did not show on the Marauder's Map, nor did its inhabitants, although, admittedly, this could simply be because the map's makers never found the room to plot it in the first place.

Witch or wizard users of the room were advised to be very explicit and secretive about what they required the room for. Those who did not would find that other wizards could enter the room and see what they were up to if they had an idea of how the room was being used.

The first mention of the Room of Requirement was when Harry heard Dumbledore tell of his discovery of a room full of chamber pots when he was in dire need of such facilities. Alas, he was never able to repeat his success, and find the room again, like many other wizards who happened upon the room.

The Room of Requirement wasn't the only creative use of space in the Harry Potter Universe, of course. There was also Hermione's beaded handbag. Reminiscent of the carpetbag belonging to Mary Poppins, Hermione had placed an undetectable extension charm on the handbag, inside of which she was able to keep all manner of objects, some much bigger than the bag appears to be, during the hunt for Voldemort's horcruxes.

Then there's the Weasley's tent. It may *look* like a two-person pitch at first glance, but when you entered you discovered a fully furnished canvas palace, complete with dining table, kitchen, bathroom, and bedrooms. It was such a sufficiently impressive and creative use of space that it inspired Harry to declare, "I love magic!"

So how might muggles be able to pull off the 'bigger on the inside' technology? How can science be used to creatively use the more curious aspects of the nature of space?

Muggle Theories of Gravity in Space

One possible way is using gravity. Now, Aristotle had explained the way objects fell to the ground by suggesting it was simply their natural resting place. And for solid 'earthy' bodies, that was the center of the Earth. He even suggested this was the focus of the entire universe, and that, if you moved the Earth itself, there would still be an abstract point in space that represented this gravitational focus, though of course he didn't *call* it gravity.

Newton, who *did* call it gravity, began a theory that culminated in the idea that all masses created a field of gravity, which exerted a force on any mass place within it. But, for Einstein, the gravity field is really a distortion, or warp, of the geometry of space and time. In other words, mass bends space. The matter of a body bends the space around it. And the more mass you have in one spot, the more bent that spot gets.

You can try experimenting with this idea quite easily, without even leaving your bed. Imagine space is like the surface of your duvet. At least, like your duvet when the bed is made, not crumpled up in a heap, as it normally is first thing in the morning. Okay, on this tidy, flat duvet you plonk a planet. Well, not an entire planet, but a simulation of a planet: a soccer ball. Better still, a bowling ball, if you have one to hand.

The thing is this: the heavier or more massive the ball, the bigger the dip. Not only that, but that dip around the ball seems to pull stuff nearby toward it. If you roll some smaller balls across the bed, you will see this. Now if you think of your balls as planets, and the duvet as the fabric of space, you can see just what Einstein was banging on about. That's how gravity works: mass bends space. Not only that, but next time you get bad press for lazing about in bed, you can always suggest you're simply contemplating Einstein's General Theory of Relativity. Might work.

So, our best theory of gravity is the idea that gravity is a bending of space and time. And a free object simply moves along the shortest path through space-time.

The Muggle Use of Space

Now, if muggles made a Room of Requirement out of the right kind of stuff, then they could use all this gravity-bending to build a bubble, of sorts, that's bigger on the inside than out. The stuff they would use would be a very exotic type of matter, but hey, such creative use of space was never going to be easy.

Imagine one of the tiniest spiders from the Forbidden Forest, crawling along a flat vertical wall. As can be seen, the square of the vertical wall has within it an attached lobe, which is somewhat balloon-shaped. The narrow opening in the wall acts as a throat, which opens out into a much bigger area. And if you scale this figure up into 3D, you roughly have a similar situation to the Room of Requirement, though the Weasley's tent, and Hermione's beaded handbag, may prove more of a challenge. Another challenge is that the material you'd need to use would be 'exotic matter'. It's weird stuff. If you pumped your car tires with it, they'd get flatter!

And there are other challenges, such as the question of making a space, or room, sentient. Scholars are developing room sentience. They're working on a fusion of human-computer interaction, sensor technology, and artificial intelligence, coupled with integrated indoor ecology control systems, with the aim of establishing intelligent rooms. These rooms are spaces where awareness of human activities and context situates the environment's behavior. In short, the space reacts and interacts with its human occupants. But if a room were to be truly sentient, as in sensing, thinking, and reacting to human activity and thought, then we'd probably need a room made of flesh.

And that would require a science beyond magic.

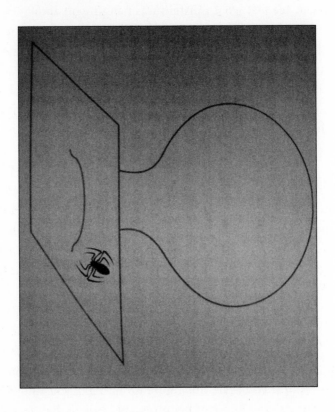

ARE FORCE FIELDS THE MUGGLE VERSION OF SHIELD CHARMS?

Picture this: Hogwarts castle under siege from all manner of death eaters and dark creatures. In an external courtyard of the castle, and under cover of darkness, Professor McGonagall declares, "Hogwarts is threatened! Man the boundaries and protect us! Do your duty to our school!" Countless castle statues and gargoyles thunder past, heading toward the viaduct and the position of the invading dark forces, all to McGonagall's evident delight, "I've always wanted to use that spell."

A few yards away, Professor Flitwick waved his wand aloft and the summoning began, "*Protego Maxima. Fianto Duri. Repello Inimicum*". As other witches and wizards join with Flitwick, we see a momentary disturbance in the dark skies above the castle. Together they conjure a magical shield, expanding ever outward, bubbling and blooming over the castle's estate, as far below the statues and gargoyles marched along the viaduct and took their posts along the perimeter.

Flitwick had conjured up a shield charm on the castle. *Protego Maxima* was a strong shield charm which, when cast in combination with *Fianto Duri* and *Repello Inimicum* raised an almost unbreakable magical wall of defense. The charms created magical barriers to deter and deflect physical objects and spells, which could work to protect a person or a locale. When done well, conjurations would often rebound directly off a shield charm, back towards the caster or ricocheting off in random space as soon as they hit the shield. Shield charms seem so easily summoned in magic. But could muggle technology conjure up something similar?

Fantasy Force Fields

All matter is made of atoms. Atoms are bound by forces. Take away atoms. Leave the forces behind. That's a force field. Well, at least that's a force field in science fantasy. In physics, a force field can have a different meaning, but for fantasy it's usually an invisible protective sphere, or wall of force. It's a concept we all know and love. It's what happens when aliens, rogue asteroids, or killer wizards are winging your way. Just slap up a force field and pour a cocktail. Job done.

The first force field found its way into fantasy in America in the 1930's and 40's. In E.E. 'Doc' Smith's *Skylark* and *Lensman* books, when force fields were under attack they typically glowed red and orange and then all the way through the spectrum until they got to violet and black, at which point they broke down. Smith's fields were a forerunner of the modern deflector on the S.S. Enterprise in *Star Trek*.

In modern movies, force fields are accepted without question. Take the 1996 movie, *Independence Day*, for example. Disabling the protective force fields of the spaceships of the invading forces is a key plot point. It renders the hostile aliens vulnerable to nuclear attack. And yet the force field is presented seemingly without need of explanation. This situation has become so endemic in fantasy that stories commonly have force fields surrounding entire solar systems!

The truth is more problematic. There is no known force capable of repelling all objects and energies. But muggles are working on it. Scientists at NASA's Kennedy Space Center and the NASA Institute for Advanced Concepts are researching the possibility of electric shields, for use on Earth, or even bases on the moon, for example. Most of the deadly radiation in space is made up of electrically charged particles. So why not use a powerful electric field that has the same charge as the incoming radiation, thus deflecting the radiation away?

Real Life Force Fields

Recently, scientists found an invisible shield, around 7,200 miles (11,600 km) above Earth. And this shield certainly helps repel 'dark' forces. The shield blocks 'killer electrons', which would otherwise be free to bombard our little planet. Now these 'killer electrons' can whizz about the globe at near-light speed. They spook astronauts, fry satellites, and wreck space systems. And, if they hit Earth on a massive scale, they could take out power grids, radically alter our climate, and send cancer rates spiking to an all-time high.

The nature of the shield itself remains a mystery. Though scientists know of its existence through its effects, they are puzzled about its formation and function. But there is little doubt this real-life shield is somewhat like the shields created by shield charms at Hogwarts, and force fields on *Star Trek*. The main difference is that rather than repelling death eaters or aliens, this invisible shield blocks high-energy electrons.

7,200 miles above the Earth puts this invisible shield in the Van Allen radiation belts. The Van Allen belts are two doughnut-shaped rings that sit in the Earth's atmosphere. They're replete with high-energy protons and electrons, and held in sway by the Earth's magnetic field. They ebb and flow, and shrink and swell, in response to inbound energy surges from our sun. The Van Allen belts were found in 1958, and are composed of an inner and outer belt extending up to 25,000 miles (40,000km) above the planet's surface.

But recently a third, transient 'storage ring' was found, hidden in the belts. This mystery third ring was discovered by the twin Van Allen Probes, launched in 2012. The ring seems to wax and wane with the intensity of space weather. But it blocks the ultrafast electrons from breaching its barrier, and probing down deeper towards the Earth. Science teams are a little puzzled by the phenomenon, but hope to learn from nature's example, which may help us understand how to make artificial force fields.

The third ring is like a glass wall in space, akin to that bubble in the dark skies above Hogwarts. At first, scientists were worried that the highly-charged electrons, which are looping about the globe at speeds greater than 100,000 miles per second, would veer down into the upper atmosphere. But the 'glass wall' stops the electrons before they get that far. If the science teams can work out how the glass wall operates, they may be able to mimic its capabilities and make a man-made barrier that does the same.

Scientists have been working out how such a glass wall might be created and sustained in space. One theory is that the wall is made by the Earth's magnetic field lines. These magnetic lines capture and control charged particles, such as protons and electrons. The particles are made to prance between the planet's poles like angry birds on a wire. Another theory is that radio signals from human activity on Earth could be scattering the charged electrons at the barrier, preventing their downward motion. But neither theory holds scientific water. At the moment, the science teams simply don't know how the slow and steady motion of these particles can suddenly conspire to create such a sharp and stubborn boundary at this 'glass wall' in space.

And yet a third theory is more interesting still. It suggests that a giant cloud of cold, electrically charged gas, beginning about 600 miles (960km) above the Earth, is scattering the electrons at the boundary. The name of this giant cloud? The plasma sphere! Now, that *does* sound like a force field.

COULD AGE LINES
EVER BECOME A REAL
TECHNOLOGY?

The age line was an enchantment that prevented anyone with the incorrect age from accessing certain objects or areas. It was used in *Harry Potter and the Goblet of Fire* to prevent underage voting. Of course, in muggle society, children are also limited in this way as they can't vote until the age of eighteen in most countries.

There are many situations in society where there is reason to limit someone's freedom based on their age, such as driving, drinking, working, and even playing. Each situation has different reasons for the age limit, whether for safety reasons, to provide fair competition, or to ensure that a person has gathered a reasonable level of accountability and life experience. The age line basically allows a person's real age to be used to discriminate against them, which sounds a bit alarming when put that way. Will we ever develop a technology like this or is it here already?

Age Restrictions

When Dumbledore drew the age line, it was to help enforce age restrictions imposed by the Ministry of Magic. The ministry had determined that the extremely dangerous Triwizard tournament should not be entered by anyone under the age of seventeen due to the mortal risks involved. As such, the age line was used to prevent anyone under that age from getting close enough to enter their name into the goblet of fire. In the wizarding world, seventeen is a significant age because wizards are not allowed to use magic outside of school until that age.

In muggle society, age restrictions are an everyday thing. In the UK, you can't work full-time until you're sixteen, and to get a credit card, drive, or vote (except in Scotland), you need to be at least eighteen years of age. Eighteen is significant in many countries because it represents what's known as the age of majority. This is when a person is considered accountable for their own actions and their parents no longer have legal control over their affairs.

Societies introduce age restrictions for good reason, although younger citizens may not agree. For example, the age appropriateness of a movie is denoted by a rating system (PG, PG13, R, etc.) This is to protect younger viewers from any adverse psychological responses to the movie. It's actually not an offence for a younger person to watch an age-restricted movie outside of a licensed cinema, however there is a responsibility for adults to ensure that younger viewers don't suffer any ill-effects due to viewing a film. As such, if staff at a cinema are in doubt about a person's age, they are legally required to ask for proof of age in the form of an official I.D. that includes a photo and date of birth such as a passport or driver's license.

The same applies at premises with a license to sell alcohol; if a person can't prove they are old enough, then they are refused service by the staff. Many drinking establishments also employ bouncers or security staff at the door who provide a physical barrier to entry. In essence, they have a similar function to an age line and are the first line of defense against age deception.

Age and Deception

"An Age Line! ... Well, that should be fooled by an Ageing Potion, shouldn't it? And once your name's in that Goblet, you're laughing—it can't tell whether you're seventeen or not!" —Fred Weasley, *Harry Potter and the Goblet of Fire*

Youngsters find ingenious ways to try and deceive the powers that be, but their degree of deception need only be as complicated as the security measures put in place to apprehend them. When it comes to face-to-face deception, beyond just barefaced fibbing, a youngster might resort to behaving or dressing in a way that makes them appear older. If this doesn't work, they might borrow an older person's I.D. card or even obtain a counterfeit I.D. However, relatively recently the increasing use of online consumerism and social media has necessitated new requirements for age verification—often accompanied by different methods to circumvent them. For example, many social media websites, such as Twitter, Facebook, Snapchat and Instagram have a minimum age restriction of thirteen. YouTube's age limit is also thirteen, but only if the youngster has their parents' permission; otherwise, it's actually eighteen. Despite these age restrictions, more than half of British youngsters under the age of thirteen still have a profile on at least one social networking site.

A reason this is possible is that age verification on many of the websites make use of what's called an age affirmation page, where the user simply enters a date of birth or clicks a checkbox to declare that they are over 18. So, a youngster can just invent a suitable age when setting up an account and the website moderators will be none the wiser. It's almost like the Weasley brothers walking up to the age line and simply saying they are above seventeen to gain access.

However, when the Weasley brothers and friends took the ageing potion to try and fool the age line by increasing their biological age, the subterfuge didn't work. It seems the age line didn't need to make use of their apparent biological age. Instead, it was able to discern their chronological age i.e. the days and years that have passed since the date of their birth.

In most muggle societies, age verification is typically done by humans, but the introduction of increasingly sophisticated technologies has been steadily moving this responsibility over to computer systems and software. When it comes to online businesses, there are various age verification methods available. This includes the classic checking of identity documents in person, as well as credit card verification and online identity check software.

One online source describes its identity verification solutions as "calling upon an extensive wealth of consumer data, including one of the largest dates of birth files in the UK." So, anytime a user provides an indication of who they are, such as through a password login or fingerprint, their identity can be cross referenced against a stored database somewhere that contains information about their age. As long as that information is reliable, then their chronological age can be correctly verified. But what if there's no available record of a person's chronological age? Is there anything within our biology that could allow our age to be determined?

Age Discrimination

As we age, different cells in our body deteriorate, reducing their ability to function normally. Some organs also experience a decrease in the number of cells due to the cells dying off and not being replaced. These are some of the causes of biological ageing, also known as senescence from the Latin word senescere meaning "to grow old." Common physiological changes include less elastic skin, weaker eyesight, and a reduction in the range of frequencies we can hear.

There are already technologies that have made use of some of these biological changes to enforce restrictions based on age. One example is the Mosquito anti-loitering device that is described by one online retailer as "The most effective tool for dispersing groups of misbehaving teenagers." The device works by exploiting the fact that as we age, our ears become less sensitive to higher frequency sounds. This is known as presbycusis. As such, the designers of the Mosquito set the unit to emit an irritating tone at a frequency of 17 Kilohertz, which they say is inaudible to the majority of people over the age of 25.

Although the high frequency tone is harmless, it can be so annoying that those who can hear it generally prefer to avoid the area surrounding the device. Its effects can be felt up to 40 meters away, but it doesn't penetrate solid walls.

In this way the Mosquito acts as a rudimentary form of age line, if only in the sense that it can effectively repel people under a certain age. However, not everyone over the age of 25 is immune to its effects, and it doesn't actually present a physical barrier to entry like a doorman would. It's not certain whether the age line is meant to present a physical barrier, or function in some other way. The Mosquito alarm is an example of a way to restrict someone based on their biological age, but there are technologies that can determine someone's chronological age through their biology.

Biological Age Markers

The scientific study of biological ageing is called biogerontology, and researchers in this field have developed a variety of techniques to ascertain someone's age. One such method involves measuring the length of telomeres, which are regions on the end of chromosomes that, amongst other things, affect how quickly cells age and die. At our birth, they can be more than 10,000 base pairs long, but they lose a bit every time the cell divides, so that by old age, they can be just 4,000 base pairs long.

Assessment of telomere length can already be done in a lab, although it requires very particular circumstances and equipment such as microscopes and the collection of cell samples from the person under investigation. However, *if* an age line could somehow tap into a person's cells, it could potentially use an analysis of the telomere lengths to gain an idea of the person's biological age.

Another method that can be used to pinpoint the biologically older parts of the human body is called DNA methylation (DNAm). DNAm is a mechanism used by cells to control gene expression. The older a person is, the less the DNAm mechanism functions. This allows it to be used as a biomarker of ageing, i.e., it can be used to predict a person's biological age. This method is responsible for the so-called DNA methylation clock also called the epigenetic clock, which is believed to provide a relatively good prediction of biological age.

Using this method, a team of Dutch researchers have managed to predict a person's real age to within four years using blood samples and to within 5 years using teeth samples. However, the younger the person, the better the error rate, leading to an accuracy of about two years in younger people.

So, Could Age Lines Ever Become a Reality?

Well, for a start, the general need to protect the safety of youngsters in our societies is a strong pull toward age-restricting technology. And considering the vast number of other situations in which it is necessary to limit access to some services, it would seem that any technology that could replicate something approximating an age line would be of great value to some commercial companies. After all, commercial value is a powerful motivation to fund technology.

There do exist ways to determine age, but they mostly require samples to be taken and then time to analyze them. Whether this could somehow be done at a distance of three meters, the radius of the age line, is currently unlikely. Maybe one day a few of these methods could be combined with a technique to ascertain personal information about somebody from a distance, but as yet it doesn't look promising. We'll have to just wait and see.

WILL SOCIETY DEVELOP ARITHMANCY?

Some might say Hermione Granger simply couldn't make her mind up about the future. She clearly abhorred the school subject of divination, the art of divining the future, or foretelling future events, by various dubious tools and rituals. And yet, she also said that Arithmancy was her favorite subject. Arithmancy was also a magical discipline that studied the future, but the difference between arithmancy and divination was that arithmancy applied a more precise and mathematical approach to predicting the future, which the rationally minded Hermione preferred. (One of her many complaints about divination was that it seemed to be "a lot of guesswork.")

Arithmancy was all about the magical properties of numbers. And, as predicting the future with numbers was prominent in arithmancy, it had a little in common with the muggle practice of numerology, where people place faith in number patterns and draw pseudo-scientific conclusions from them. Arithmancy was an elective subject at Hogwarts, offered only from the third year on. Students were set essay assignments that involved the consulting and/or composition of complex number charts. Those who practiced arithmancy were called arithmancers.

Among the notable arithmancers in wizard history were Bartemius Crouch Jr., the death eater who made Voldemort's return inevitable, and Hermione herself. She became a high-ranking official in the department of magical law enforcement. So, it looks as though Harry indulged in a spot of fortune telling himself. He'd bought Hermione a copy of the book, *New Theory of Numerology*, for Christmas while they were still in school. But can numbers really divine the future?

Numbers Unlock the Cosmos

The link between numbers and nature had been recognized early by the Pythagoreans when the Pythagorean brotherhood of ancient Greece had realized that numbers were the key to comprehending the entire cosmos. The brotherhood tried to synthesize a holistic view of the universe, one which incorporated religion with science, medicine with cosmology, and mathematics with music; mind, body, and spirit as one. The very word, "philosophy" is Pythagorean in origin. When muggles use the word harmony in its wider sense, when numbers are called figures, etc., muggles speak the tongue of the brotherhood. And their approach was epoch-making; through their application of mathematics to the human experience, the Pythagoreans were founders of what the world understands today as science.

The Pythagorean brotherhood was founded in the 6th century BC. And the Pythagoreans were big on the magic of numbers. To them, philosophy was the highest music. And the highest form of philosophy was concerned with numbers, for ultimately all things are numbers. So rather than numbers leading to a reduction of human experience, it was an enrichment. The Pythagorean concept of harmony was typical of the way in which the brotherhood synthesized an interconnected view of the universe. Numbers were not tossed into the world at random. They were arranged, or arranged themselves, like the structure of crystals or a musical scale, according to the universal laws of harmony.

The basic Pythagorean notion of armonia regarded the human frame and body, too, as a kind of musical instrument. Each string within must have the right tension, the correct balance, for the patient's soul to be in tune. The musical metaphors that are still applied to medicine, such as tone and tonic and well-tempered are also part of the Pythagorean heritage.

Legend has it that Pythagoras found the link between music and math from a blacksmith. One day, Pythagoras was passing a smith at work. Hearing the sweet sound of the blacksmith striking the anvil, Pythagoras realized that such harmony must bear some relation in mathematics. He spent some time with the smith, examining the tools and exploring the simple ratios between tools and tones.

The Pythagoreans took numbers so seriously that they were prepared to kill for them. Tragically, one of nature's number secrets helped bring about the end of the brotherhood. For the Pythagoreans discovered irrational numbers. These numbers, such as $\sqrt{2}$ or π, are numbers that cannot be written down as the ratio of two integers, two whole numbers. For philosophers who believed that all of nature could be understood by number-series and number-ratios, this was a major blow.

The proof of the existence of irrational numbers is attributed to a member of the brotherhood, Hippasus of Metapontum. He is thought to have found them while thinking about the geometry of the pentagram, used by the Pythagoreans as a symbol of recognition among members and as a mark of inner health. At first, other members tried to disprove the existence of irrational numbers through logic. They failed, and today we know that almost all real numbers are irrational. Believing in the absoluteness of numbers, the Pythagoreans kept the discovery a secret, dubbing the irrational numbers *arrhetos*, unspeakable. But Hippasus let the scandal leak, and legend has it he was put to death by drowning,

"It is told that those who first brought out the irrationals from concealment into the open perished in a shipwreck, to a man. For the unutterable and the formless must needs be concealed. And those who uncovered and touched this image of life were instantly destroyed and shall remain forever exposed to the play of the eternal waves." —Mark Brake, *Revolution in Science*

Greek drama, indeed.

The Dark Archives

The modern use of numbers is no less dramatic. In the legendary *Foundation* fantasy trilogy by Isaac Asimov, a mathematics professor named Hari Seldon predicts the future using what Asimov calls psychohistory. Mathematics is used to model the past to help anticipate what will happen next, including the fall of the galactic empire. It may seem like science fiction, but a new field does something similar.

Cliodynamics (named after Clio, the Greek muse of history) claims to enable its scholars to analyze history in the hopes of finding patterns they can then use to map out the future. One of the big questions they aim to answer is, Why do civilizations collapse? Their holistic approach isn't too far removed from the Empire-mapping forecasts set out by Hari Seldon in Asimov's *Foundation*. The technique predicts a wave of widespread violence around 2020, including riots and terrorism.

This mushrooming field of cliodynamics uses the dark archives. These are data banks from the distant past, including historical documents that have only recently come online. The methods include common statistical techniques, such as spectrum analysis, on historical digitized newspapers and public records. So cliodynamics quantifies the past and makes extrapolations based on those data trends.

Experts in the field have found a pattern of social unrest. It applies to many civilizations, including dynastic China, ancient Rome, medieval England, France, Russia, and even the United States. Analysis clearly shows one-hundred-year waves of instability. And, superimposed on each wave is an additional fifty-year cycle of widespread political violence. China seems to escape the fifty-year cycles of violence, but the United States does not.

These cycles of violence have social inequality at their root. Discontent builds up over a period of time until the pressure is violently released. Scholars have mapped the way that social inequality creeps up over the decades, so much so that a breaking point is reached. A little late, reforms are finally made. But, over time, those reforms are reversed, and society lurches back to a state of heightening social inequality. Sound familiar?

The severity of the violent spikes depends on how governments cater with the crisis. For example, the United States was in a pre-revolutionary crisis in the 1910s, but a sheer drop in violence followed due to a more progressive political era. The ruling classes made calls to reign in corporations, and allowed workers vital reforms. Such policies reduced the pressure, and prevented revolution. Likewise, 19th century Britain was able to avoid the kind of violent revolution that happened in France by making amelidratory reforms. However, the usual way for the cycle to resolve itself is through violence.

Much has been made of the dark archives. Cliodynamics shows that muggles can find much value in the reams of non-digitized records that most don't even know contain prophetic treasure. The world is taking a closer look. And historians are starting to work with mathematicians to embrace the new techniques. Our future may soon be divined from our past.

COULD MUGGLES MASTER THE MEMORY TRANSFER OF A PENSIEVE?

L egend had it that the Pensieve was older than Hogwarts itself. The founders of the school were said to have discovered the Pensieve half-buried in the ground, made of an ancient stone, and carved with a strange sort of Saxon runes. The discovery not only placed the Pensieve's creation before the founding of Hogwarts, but was also rumored to be one of the reasons the school was set in such a remote location.

The Pensieve was an artifact used to store and sort memories. Looking much like a shallow stone or metal basin, the Pensieve, inlaid with carvings of the runes and strange symbols, sat filled with a silvery substance, some-what like the metal mercury in appearance, and yet light and cloud-like. The wizards who used the Pensieve would transfer a wispy form of their thoughts and recollections, from mind to artifact, by way of a wand. A Pensieve was the sum of the collected memories of all the witches and wizards who had siphoned their thoughts into it.

An illustrious line of Hogwarts headmasters and headmistresses left behind their legacy, in the form of memory. The collection represented a treasured library of reference for any witch or wizard headed for the post in the future. Presumably, Dumbledore himself had added his memories to the mix, not the least of which were his recollections of the rise and fall of Voldemort. Dumbledore once remarked that he found the Pensieve invaluable in sorting through the mind, spotting links and patterns that might otherwise be missed.

The word 'Pensieve' is portmanteau. It's a combination of the words pensive and sieve. To sieve, of course, is to drain, sort, or separate. And pensive, derived from French and originally from the Latin *pensare*, means to ponder, and in English also means thoughtful or reflective. Together, it can be seen that Pensieve means exactly what Dumbledore used it for—a relic that allows for the sorting of thoughts and memories.

A note of caution was stressed on the use of the Pensieve. As was seen when Dumbledore used Harry as wizard witness to past events, memories could be viewed from a non-participant, third-person perspective. And, given the possible highly intimate and personal nature of stored memories, the Pensieve was open to potential abuse. So, most Pensieves were entombed with their owners, along with the memories they held. Some witches and wizards passed on their Pensieve and memories to other witches and wizards, as was the case with the Hogwarts Pensieve. But, what advances have there been in muggle memory transfer? And are we anywhere near developing the tech that looks anything like a Pensieve?

Memory Edits

Manipulation of memory has a long history in fantasy. *The Memory Clearing House*, a story written way back in 1892 by Israel Zangwill, tells the tale of both removal and borrowing of memories. As with the Pensieve, memory removal is done through a device, a noemagraph, or thought-writer, which receives the impression of thought on a sensitized plate that acts as a medium between minds. In John Brunner's *The Long Result*, written in 1965, a device is used to time-lock sensitive memories, which can then be recalled by the protagonist many years later.

Perhaps the best-known memory edits appear in the movies. In the *Men in Black* movie franchise, MiB agents use a neuralyzer, a device looking much like a normal cigar tube that, when flashed, erases in the recipient the memories of the past hours, days, weeks, months, or years, depending on the selected settings. In the 1998 movie, *Dark City*, aliens are able to memory-edit humans on a mass basis. False memories of vacations on Mars are implanted in *Total Recall*, based on the Phillip K. Dick story, *We Can Remember It for You Wholesale*, written in 1966. The implanted memories are cheaper than actually traveling to Mars, but the trouble arises when the embedded memories get entangled with reality. Memory intrusion also features in the movie *Inception*, though in this case it is without the subject's permission and via a dream-hacking technology. Little wonder muggles are fascinated by a future where memory might be engineered.

Shakespeare in a Syringe

Between the late 1950s and the mid-1970s, it started to look like memory editing might actually be possible. The tale began with the chemical transfer of memory in animals. It looked as though memories were stored in chemicals, which could be moved from creature to creature. The research was exciting. If memories truly were programmed into molecules, what else might be possible? Kindergarten kids could master multiplication by swallowing a pill, perhaps. College students could become fluent in a foreign language by having it inserted under their skin. And actors could memorize *The Complete Works of Shakespeare* by essentially having the great bard injected into their bloodstreams.

The memory engineering process began well. Muggle scientists thought they had taken substances corresponding to memories, out of the brain of one animal and into another, with beneficial results. Say the first creature had been trained in some task. After memory transfer, it seemed that the second creature would know how to perform the same task, but with far less training. The second creature essentially had a head start advantage, based on the memory edit.

It all began with worms. The first tests were done on planarian worms, a form of flatworm. They were trained to scrunch up their bodies in reaction to light. If the scientists shone a light on the worms as they wended their way across the bottom of a shallow water tank, and at the same time gave them a mild electric shock, in time the worms would learn to associate the light with the shock. Eventually, the worms would scrunch up their bodies whenever the light was shone, whether or not a shock was also dealt. Worms that scrunched in reaction to light alone were considered expert worms. And yet novice worms, who had no such water tank experience, were seen to behave in the same way as the expert worms, once tissue had been transferred from expert worms to novice ones.

In the mid-1960s, the memory-edit work moved onto mammals. A scientific controversy had already raged about the interpretation of the worm experiments. Now, the stakes were even higher. And yet, decades worth of work was disregarded when a single paper was published in 1964. It came from the influential laboratory of Nobel Laureate Melvin Calvin. The paper described work on planarian worms and undermined all the early research on the chemical transfer of memory.

By 1972, the field was almost dead. A five-page report in favor of the idea of the chemical transfer of memory was published in *Nature*, the most prestigious journal for biology. And yet, accompanying it was a fifteen-page critical editorial, which significantly weakened the credibility of the field of memory edits, but there are no published reports that decisively disprove the idea of the transfer of memory. Many of the positive results from the early research have never been explained away. Memory transfer was never disproved. It seems scientists simply grew tired of it, or more interesting topics came along. And, subsequently, *Nature* has used critical editorials to the disadvantage of controversial science in other fields.

CAN MUGGLES DEVELOP THEIR OWN FORM OF TELEPORTATION?

Wizards pop up in the weirdest places, don't they? One moment they're contemplating cauldrons in Diagon Alley, and the next they're sipping a butterbeer at The Three Broomsticks. But then, getting about in the wizarding world is easy. There are so many instant travel options, such as brooms, Floo Powder and portkeys.

Maybe the most fascinating method of travel is apparition. This magical method of transportation is all about the three D's: destination, determination, and deliberation. A travelling wizard must be determinedly focused on their desired destination, move with haste, disappearing from their current location and instantly reappearing at the desired location, but with deliberation. In short, apparition is a form of teleportation.

And, like future teleportation, its speed and ease of travel is somewhat balanced by the drawbacks. Not only does apparition come with a noise, ranging from a quiet pop to a loud crack, but it can lead to injury if botched. Whereas even house-elves can apparate, and skilled wizards and witches perform apparition without a wand, novice wizards can drop body parts when they practice apparating. This occurs when the wizard has insufficient determination to reach their goal. Some body parts simply fail to arrive at the destination with the wizard.

The other wizarding phenomenon similar to teleportation is the use of the vanishing cabinets. A pair of vanishing cabinets acts as a passageway betwixt two places. An object placed in one cabinet will appear in the other. They can transport wizards too. They were very popular during the first wizarding war, when, to hide from a death eater attack, a wizard would disappear to the other cabinet, until the peril had passed. But if one of the cabinets is broken, an object travelling between the two is trapped in a type of limbo.

So, how possible is teleportation? And what's the lowdown on the topic in fantasy and fact?

A Brief History of Fantasy Teleportation

Teleportation has been a fantasy classic for many years. It's the dream of being able to transmit matter across space, in an instant, and recreate it exactly, in another location. The notion is mentioned in an early Jewish myth, where it's referred to as *Kefitzat Haderech*, which literally means "the shortening of the way," or "short cut".

This mythical term was adapted by American fantasy writer, Frank Herbert. In his 1965 novel, *Dune*, often cited as the world's best-selling sci-fi novel, Herbert refers to the hero of the book as the *Kwisatz Haderach*, a genetically-engineered short cut to a human future and the creation of a *homo superior*.

Many myths and magical tales have people being spirited away, as if by teleportation. But these episodes are usually portrayed as mystical or divine gifts. The first exploration of modern teleportation came with the 1877 tale, *The Man Without a Body*, by Edward Page. In this short story, the scientist hero, after successfully disassembling his cat, telegraphs its atoms and then reassembles them. Alas, while trying to repeat the experiment on himself, a luckless, and badly timed, power cut meant only his head was transmitted.

This kind of teleportation mishap seems common. In *The Fly*, originally a 1957 short story but also made into three movies, the grotesque consequences of teleportation are explored in detail. A scientist, experimenting with teleportation, accidentally ends up fusing himself with a fly in what is meant to be a harrowing, but is in fact largely hilarious, sequence of events.

Over the years, teleportation has become a staple in fantasy tales. In the 1939 *Buck Rogers* serial, teleportation was the travel means of choice for moving characters from one place to another. Most people, of course, associate teleportation with the phrase, "Beam me up, Scotty", from Star Trek. In this influential television series, it had been originally planned to have the characters land their spacecraft on planetary surfaces. But budgetary constraints on the special effects department meant a more creative solution was needed. The transporter was born.

And teleportation is fused with time travel in the *Terminator* series of movies, as well as the *Stargate* franchise, in which wormholes are used to help transfer from one space-time to another.

A Brief History of Muggle Teleportation

Teleportation in fantasy is often described using the language of quantum technology. Matter transmission of this kind would mean the original object or person being destroyed, and pieced together elsewhere. You can see this method might have attendant difficulties! What if the 'piecing together' doesn't go according to plan? There are trillions of atoms in the human body. And that means a person would have to be broken down into individual ones before each were catalogued, digitized, and teleported. Then the whole process would have to be done in reverse, to assemble them in the new location. And where would the soul go, assuming such a thing exists. (This also raises the possibility of perhaps splitting the soul into constituent parts, though that's another story.)

One way of avoiding the problem of piecemeal teleportation is duplication. In this schema, rather than the simultaneous destruction and recreation of the object or person, the teleportation simply generates an exact duplicate at distance. But then, this method has the different problem as to exactly who is the 'original'.

Muggles have been slow to play catch-up on the fantasy. In 2002, Australian scientists successfully teleported a laser beam by scanning a specific photon, copying it and then recreating it at an arbitrary distance. Science teams in Germany and America then independently teleported ions of calcium and beryllium using very similar technique to one another. A further development occurred in Denmark in 2006. Here, scientists effectively teleported an object half a meter. Although miniscule in scale, the object in the Danish experiment was nonetheless constructed from billions of atoms.

Professor Michio Kaku of City University of New York, believes the technology to teleport a living person elsewhere on Earth, or even to outer space, could be available by the end of the century. Professor Kaku, known as a noted futurist who admits to optimism on the topics of real time travel and invisibility, has made a study of various fantasy technologies and determined some will one day happen.

Professor Kaku suggested, "You know the expression beam me up Scotty, we used to laugh at it. We physicists used to laugh when someone talked about teleportation and invisibility, something like that, but we don't laugh anymore—we realized we were wrong on this one. Quantum teleportation already exists. At an atomic level, we do it already. It's called quantum entanglement". He described the process as one that allows connections, somewhat like an umbilical cord, to be formed between atoms, with their information transmitted between others further away. "I think within a decade we will teleport the first molecule," Professor Kaku concluded.

WHAT TECHNOLOGY IS THE MUGGLE VERSION OF THE HORCRUX?

It was Voldemort's party piece. He split his soul, and hid fragments of it in objects outside his body. Then, even if his body was attacked or destroyed, he could not die, as part of his soul was transcended; it stayed earthbound and undamaged. The word used for a magical object in which a person has concealed part of their soul in this way is a horcrux. And it's a fantastic vehicle of transcendence.

To create a horcrux, a wizard had to commit calculated murder. The act of murder would damage the soul of the wizard. And that damage could be used to cast a spell, which would rip off a damaged fragment of the soul and encase it in an object. If the wizard were later killed, he would transcend death, living on in a non-corporeal form. But there were also ways of re-possessing a physical body.

Voldemort pushed his soul to the limit in creating his seven horcruxes. It rendered his soul volatile, and liable to break apart if he was killed. The process began when he created a horcrux during what was his fifth year at Hogwarts. Horcruxes were originally thought of as being singular, but Voldemort created seven, presumably in the hope that seven would make him stronger than just one.

In fact, as in fantasy, humans have long sought to transcend our reality into some kind of new existence. Historically, the idea of transcendence has mostly been within the domain of faith. But, increasingly, the notion grew that science may somehow be able to move us beyond our physical limitations. The technology of transcendence may be just around the corner.

Transcendence and Elec-trickery

Einstein once remarked, "Reality is merely an illusion, albeit a very persistent one." When a fascination with electricity crackled into life, the notion of transcendence swiftly sparked into being. Benjamin Franklin had brought electricity down to Earth through the lightning conductor. Michael Faraday conjured a cocktail of electricity and magnetism in the dynamo. And the exposed sciatic nerve of a frog's leg kicked Luigi Galvani into the discovery of bioelectricity.

Yet some felt there was darker magic still. The first science to materialize after Newton, electricity had a long and legendary history. From ancient times, people had treasured the doctrine of affinities. The attraction of amber illustrated the very idea of virtue dwelling within a special substance. As if through magic, the magnet's enchanting property of virtue was bestowed on other objects through touch alone. The possibilities seemed endless.

Into this climate came Mary Shelley and her epoch-making *Frankenstein*. In June 1816, Mary Wollstonecraft Godwin and her intended, Percy Bysshe Shelley, had visited Lord Byron at Lake Geneva. That year without a summer, it had seemed the entire planet was frozen in a volcanic winter triggered by the eruption of Mount Tambora the previous year. Kept indoors by the incessant weather, the romantics turned to conversation. Their reading matter was mostly fantasy, including *Fantasmagoriana*, an anthology of German ghost stories that talked of animating life. One experiment credited to Erasmus Darwin reported that a piece of vermicelli, preserved in a glass case, had begun to move with voluntary motion. Mary later fell into a waking nightmare in which a "pale student of unhallowed arts kneeling beside the thing he had put together," haunted her. The nightmare was the seed of Frankenstein.

Frankenstein is essentially a tale of transcendence. Victor Frankenstein is the Faust of the New Philosophy. The subtitle of the novel, *The Modern Prometheus*, also compares with Prometheus's theft of fire from the gods for profit. Victor's dream is unlimited power through science, a power brought about by human, not supernatural, agency. Victor rejects the dark arts of old world alchemists Paracelsus, Albertus Magnus, and Cornelius Agrippa, and turns to face the future. He becomes obsessed with the essence of life. He manages to unravel the agency through which dead matter may be given the vital spring of life. Intending his creation to be beautiful, Victor builds a mechanically sound but grotesque creature using cadaver spares from charnel-houses. Only when inspired by the new unbridled science, is he gifted this terrible triumph of creation.

With the story of *Frankenstein*, Mary Shelley tapped into the excitement about electricity at the time. Throughout Europe there was a thrill about the use of this new force, and feverish research on the potential of electricity to sustain, create, and even transcend life itself. The potency of this new power is evident in Victor Frankenstein's words about the new science and its wizards, "They ascend into the heavens: they have discovered how the blood circulates, and the nature of the air that we breathe. They have acquired new and almost unlimited powers; they can command the thunders of heaven, mimic the earthquake, and even mock the invisible world with its shadows."

Darwin Among the Machines

The first promise of electricity was never realized. But the new technology of transcendence is machine intelligence. Darwin's theories of evolution were first taken into the machine world by British novelist Samuel Butler. In his 1872 book, *Erewhon* (an anagram of the word nowhere), the hero travels to a future society that has banned technological evolution. They feared that machines would evolve, become intelligent and aware, and enslave their human masters.

"Complex now, but how much simpler and more intelligibly organized may it not become in another 100,000 years? Or in 20,000? For man at present believes that his interest lies in that direction; he spends an incalculable amount of labor and time and thought in making machines breed always better and better; he has already succeeded in effecting much that at one time appeared impossible, and there seem no limits to the results of accumulated improvements if they are allowed to descend with modification from generation to generation." —Samuel Butler, *Erewhon*

The perils of allowing machines to think were therefore present from the very early days of fantasy fiction. And yet, scholars now think that humans might actually *become* the machines. Once artificial intelligence is advanced enough, engineers will be able to upload human consciousness to a machine. The assumption is that consciousness can somehow be replicated into a series of brain emulations, and so the person concerned can be encapsulated in the same way a soul would be encased within a horcrux.

If this happened to you, you would be transcended. Your new sleeve or body could be robot or android, or you could simply live in virtual reality. You would think and act a thousand times faster, and be very much fitter for the future. In fact, like Voldemort, why stop at one? If your consciousness can be uploaded, why not decant yourself into seven different sleeves, in the same way Voldemort created his seven horcruxes?

How would it feel, to be transcended? Perhaps we can try answer that question by considering *Mary Shelley's Frankenstein*, the 1994 movie, directed by Kenneth Branagh, starring Robert De Niro. Unlike the novel, Brannagh has Victor Frankenstein reanimate the murdered Elizabeth, the love of his life. It is a strangely moving, yet disturbing and grotesque sequence in the film where Frankenstein's reanimated dead love realizes the unnatural and utterly horrific thing he's done. The reanimated Elizabeth goes berserk, alienated by the monstrous state of limbo in which she finds herself. Maybe this is what it would be like, to be transcended.

IS IT JUST WIZARDS AND WITCHES WHO WAVE WANDS?

Redwood, rosewood, rowan, spruce, or vine. Blackthorn, beech, or willow. In the Harry Potter Universe, wand wood came in almost forty different varieties. And wand makers also catered to a cornucopia of different cores. You could fill your wand with phoenix feather, dragon heartstring, unicorn hair, or thirteen other exotic options. The wand was the wizard's weapon of choice. It was the object through which a wizard or witch channeled their magic. The wands were made from these various woods, with the magical substance running through their core, and were of varying lengths and pliability for more focused and wide-ranging results.

Indeed, in the magical world, most spells were done with wands, for wandless magic took more skill and acumen. Wand magic was commonly conjured up with an incantation. But wiser and more experienced wizards could also cast nonverbal spells, concealing the spell until cast, and possibly preventing an opponent from protecting against it.

Every single wand was unique. Though wand cores may come from the same creature, and the wood from the same tree, no two wands were exactly alike. And, depending on the character of the wood and magical creature from which it came, wands were said to be quasi-sentient; being imbued with a great deal of magic made them somewhat sentient and animate. Wands were made and sold in Great Britain by the Ollivander family, who began wand manufacture in 382 BC.

The study of the history and the magical properties of wands was known as wandlore. Wandlore was considered a complex and mysterious branch of magic, which included the idea that a wand chose the wizard, and not the other way around, and that wands could switch allegiance. But is it just witches and wizards who wave wands, or does wandlore have a much longer history?

Wand as Symbol

These days, it seems the wand is everywhere. Most famously, perhaps, are wands wielded by many fictional characters such as by Cinderella's godmother, by the mages and warlocks of the massively multiplayer online role-playing game, *World of Warcraft*, and in Tolkien's *The Hobbit*, by Gandalf, whose very name in Northern Mannish (one of Tolkien's invented languages) means "elf of the wand."

Science and tech have adopted the word wand for many inventions. Wand can refer colloquially to a handheld metal detector, such as at airports and high security buildings. Wand is also used to mean the steering wheel control stalks for lights, windscreen and the like, and in music, the word wand applies to the modern model of a conductor's baton, used for conjuring music out of an ensemble of musicians.

Magic wands have been with us for millennia. Some Stone Age cave paintings depict early humans holding wands, which may have been symbolic representations of their power. Wands also appear in the artwork of the ancient Egyptians. Indeed, the dating of the Ollivander family wand manufacture to 382 BC is probably in recognition of the evidence of the above, and the knowledge that matching wand, wood, and wizard was a facet of the Druidic cultures that existed in Europe prior to Christianity. The wizards, or sorcerers, of the Druid magical ceremonies would wield wands made of willow, yew, hawthorn, or other tree woods they held to be hallowed. Such wands were only carved at dusk or dawn, as this was thought to be the best time to capture the sun's power. And the carving itself was carried out using a sacred knife, which had been dipped in blood.

The wand as a symbol of power makes an appearance in Christianity, too. One Old Testament tale of Moses has him wield a magic wand in the form of a shepherd's staff to both divide the Red Sea and to draw water out of a rock. And a 4th century depiction of Jesus, restoring life to Lazarus of Bethany, shows Christ touching Lazarus with a wand, implying that the wand served as a rod through which supernatural forces could be conducted.

Like the Elder Wand, ancient magic wands are still in existence. From Egypt and dating back to 2800 BC, these ancient wands are carved from hippopotamus ivory. As the hippopotamus is known as a highly aggressive and capricious creature, ranked among the most dangerous animals in Africa, any wizard that wielded a wand made from this beast would surely benefit also from its formidable power.

The hippopotamus wasn't the only creature featured in ancient Egyptian wands. So-called apotropaic wands (apotropaic here meaning to prevent evil) were used to ward away the power of demons. Dating from around 2100 BC, they were curved and decorated with magical creatures such as the griffin and the sphinx, as well as more ordinary animals such as bulls and baboons, cats and crocodiles, panthers and lions (assuming you could catch the creatures), snakes and frogs.

The Most Mysterious Wand

The best ever wand story stems from the oldest known ceremonial burial in Western Europe. It is the year of our Lord, 1823. A lone horseman gallops through the night, decked out in tall top hat and flowing gown. His destination: the south-west coast of the land of Merlin: Wales, and in particular, the Gower peninsula. The horseman rides into history, into the past. Into *all* our pasts. The man on the horse is one of a new breed of detectives. He is Professor William Buckland, master of rocks as Professor of Geology at Oxford University, and he's armed with a hammer. Our Professor is about to make an earth-shattering discovery. He was summoned to Paviland cave, one of the caves in the Gower's limestone rocks. The cave had first been found the year before, but now in 1823, one of the world's most important finds was about to be uncovered. For what Buckland was about to unwittingly discover would help unravel the story of time itself.

In that dark cave, Buckland found the first human fossil recovered anywhere in the world. But that's not all. In Buckland's words, "I found the skeleton enveloped by a coating of a kind of ochre . . . which stained the earth, and in some parts extended itself to the distance of about half an inch around the surface of the bones . . . Close to that part of the thigh bone where the pocket is usually worn surrounded also by ochre (were) about two handfuls of the Nerita littoralis (periwinkle shells). At another part of the skeleton, viz. in contact with the ribs (were) forty or fifty fragments of ivory wands (also) some small fragments of rings made of the same ivory and found with the wands . . . Both wands and rings, as well as the Nerite shells, were stained superficially with red, and lay in the same red substance that enveloped the bones."

Buckland had also found a mammoth skull, lying along with the bones. The Professor's diagnosis was typical of his time. As a creationist, Buckland misjudged both the skeleton's age and its gender, for our Professor believed no human remains could have been older than the Biblical Great Flood. So, he wildly underestimated its true age, believing the remains to be female, mostly due to the discovery of the decorative items, including the wand, which were thought to be of elephant ivory, but are now known to be carved from the tusk of a mammoth. The mammoth wand made Buckland believe the remains belonged to a prostitute or witch. Perhaps the old Welsh witch had been based at a nearby Roman camp, but, Buckland felt it was definitely a woman. There was the wand. There was jewelery, and the remains were covered with red ochre. Indeed, to this day the human remains are still known as The Red Lady of Paviland. But this was no ordinary skeleton.

The true identity of this wanded skeleton then went on a remarkable journey. In the time since the discovery in 1823, scholars have found out some surprising things about the Red Lady. He was in fact a man, and a young man at that, no older than his twenties. Because his skeleton was found without a skull, providing estimates was something of a challenge. But he may have been about 5 ft. 8 inches tall, and around 150 pounds in weight. The absence of a skull also sadly prevented the use of forensic anthropology to recreate his face. Decapitation was common in burials across Europe in the upper Paleolithic period, though some have speculated that the skull may have washed away during a subsequent flooding of the cave. The face of this man, this wizard of the Stone Age, one of the most ancient of British ancestors, will remain a mystery.

Little did Buckland the creationist realize the true nature of his discovery. The skeleton Buckland branded a prostitute was, at the time of the discovery, the first modern human skeleton to be found anywhere in the world. It remains the oldest ever found in the UK, as well as the oldest ritual burial discovered in Western Europe. There have been further finds in the area where the Red Lady was discovered. Thousands of flints, teeth, and bones, as well as needles and bracelets have been found. Such evidence suggests that the cave was visited regularly by our human ancestors for around 10,000 years, until the last ice age would have forced them to head south.

The Red Lady is evidence of one of the oldest known wizards. It is likely the cave was sacred to Paleolithic peoples, perhaps a site for ancient pilgrimage. Shamans may have contacted the spirit world in the cave. This, along with the mammoth's skull originally found with the skeleton, have prompted suggestions that the Red Lady was himself a shaman, or at least an important tribal chieftain. Over the centuries, his wanded skeleton may have become a revered relic in this shrine-like cave.

LUMOS! HOW COULD A WAND GIVE OFF LIGHT?

In the wizarding world, there is a handy piece of magic that can turn a witch or wizard's wand into a torch without a flame. Before the invention of electric lights, owning an instrument that could be commanded to give off light at will would have been a remarkable thing indeed.

Although nowadays such an illuminating device wouldn't seem that remarkable, excepting the fact that its light is produced from the end of what is essentially a pointy stick. Many people carry around their cellphones, which have multiple functions, including instant light. There is even a function on Android phones where the user can just say *Lumos* on the Google app, to activate their phone's torch. They can also say *Nox* to turn it back off.

In this way, muggles daily brandish their cellphones as a light source in a similar way to wizard's wielding their wands. So, what would be needed to get a wand to give off light like a torch?

Lumos!

This literally enlightening charm is one of the best ways a wizard has of shedding light on dismally lit situations. Once engaged, a wizard's wand will emit light from the tip as if it were fashioned from E.T.'s finger.

When a life form uses internal reactions to create light the process is called bioluminescence. Fireflies, glow worms, angler fish, and various other animals have this ability to produce light at will but the light is often blue, green, or sometimes even red. As a wand is not a life form, we can disregard bioluminescence as the light producing mechanism.

Light can also be emitted by mixing two chemicals together. The chemicals react and give off light through a process called chemiluminescence. This is how glow sticks work. They come in a variety of colors and can be made to shine very brightly if heated, although the size of the glowing wand tip would limit the amount of light that can actually be given off. The light emission can also be stopped by making them extremely cold. So chemiluminescence is a possible candidate.

Generally, when it comes to light emission there are many possible sources but the most common involve either incandescence or luminescence. Incandescence involves light given off as a result of the temperature of matter, whereas luminescence is light given off regardless of the matter's temperature. In regard to wands this means that if *Lumos* works via incandescence then the wand should also be giving off a detectable amount of heat, whereas if it functions through luminescence then the wand could work as a much cooler light source.

Making light

Light is a form of electromagnetic (EM) radiation, transmitted as photons carrying different energies. Generally, the photons that bring us light are a result of the absorption and emission of energy by the electrons in atoms. Electrons only absorb and release energy in particular amounts, known as quanta.

Regarding luminescence, this absorbed energy can come from a variety of sources, with each type of luminescence identified by its particular energy source. For example, electroluminescence is caused by electricity, sonoluminescence is triggered by sound and photoluminescence is instigated by energy from photons.

When an electron absorbs this energy, it rises to a higher energy level, and the atom is said to be in an excited state. Straight afterward, the electron reemits that energy as a photon, causing the electron to drop back down to its original energy level, returning the atom to an unexcited or ground state. The photons emitted from the atom have different frequencies corresponding to the amount of energy they carry. The higher the photon's frequency, the more energy it has.

The frequencies of electromagnetic radiation that we can perceive with our eyes is known as visible light, encompassing the entire spectrum of colors that we can see in rainbows. Each hue or color of a rainbow corresponds to a particular frequency of visible light. Red light is the lowest frequency we can see, at roughly 430 terahertz, and the highest frequency we can see is about 770 terahertz; the frequency of blue light.

When we see light it's normally composed of a range of different frequencies, i.e., it's a conglomeration of different hues. If there is a bigger intensity in any particular frequency, then overall the light appears to gain slightly more of that hue. However, if all the frequencies of visible light are present in roughly similar proportions then overall, we perceive it as white light.

The fact that the *Lumos* charm gives off a brilliant white light indicates that the wand is giving off photons from all parts of the visible spectrum. These photons can be produced by using any of the many different forms of luminescence. Each one has its own benefits for producing light with a wand, although luminescence isn't the only available option.

Incandescence

All objects that are not at absolute zero (the coldest temperature possible, -273°C), emit photons in the form of thermal radiation. This process is known as incandescence.

If the object giving off the light is opaque, (so that most of the light coming from it is from itself, rather than reflected) then the object can be regarded as what's called a black body. In line with that, the electromagnetic (EM) radiation it gives off is considered as black body radiation.

At about 525°C, called the Draper point, most solid materials will start to visibly glow. At that temperature, the peak frequency of EM radiation is in the infrared but as some of the emission reaches into the red end of the visible spectrum, we are able to register the radiation as a dull red glow.

This is why when an incandescent light bulb is dimmed to low light (and temperature), the filament can be seen glowing red. Increasing its temperature increases the intensity of light given off (it gets brighter) as well as the peak frequency of the light. This makes its color change from red through to orange, yellow, and then white at the hottest temperatures, hence the term "white hot" for describing extreme heat.

At the temperatures that it's exposed to, the lightbulb's Tungsten filament would quickly burn out if there was oxygen present. This is why filaments are contained within a bulb that does not contain oxygen. This used to be achieved by creating a vacuum inside of the bulb, but later it was found that filling it with an inert gas like argon can slow down the evaporation of the filament, allowing it to operate at higher temperatures.

The tip of a wand could be heated until it emitted visible light, but being in an environment with oxygen means it would catch alight and burn away. This would be exacerbated by the fact that the wand is made of wood, which is far less robust than the more resilient tungsten which is used as the filament in the majority of incandescent bulbs. However, each wand does also have a core of some magical substance such as unicorn tail hair, dragon heartstring or phoenix feather. Maybe these magical substances act like filaments with properties far greater than tungsten.

Lumos Solem: Recreating Sunlight

In the movie, *Harry Potter and the Sorcerer's Stone*, Hermione uses the *lumos solem* charm to produce a strong beam of light to mimic sunlight. Could sunlight be replicated in a wand?

At their core, stars rely on thermonuclear fusion to provide the energy for photon production. Under immense pressure and temperature, atomic nuclei are fused together in the core of stars. This releases energy which gradually reaches the surface layer of the star, called the photosphere. The atoms in the photosphere absorb the energy, then release it again as mostly visible light. The actual source of visible light in a star is the photosphere, which is releasing photons as a result of its electrons being energized by radiation from within the star.

The layers beneath the photosphere are also too dense to be penetrated by visible light, meaning the star is essentially opaque beneath the photosphere. As such, a star can be considered almost like a black body and acts as an incandescent light source. In this way, although a bulb can't provide an exact light match to the sun, it still shares the property of being an incandescent light source. Could we simply put a small light bulb into the tip of a wand to provide a *Lumos* function?

Well, small incandescent light bulbs were first incorporated into torches or flashlights around 1900. They were powered by electricity from dry-cell batteries. These bulbs had limits on how long they'd last and how bright they could shine. Nowadays, incandescent bulbs can be made as small as a quarter of an inch in length, with powers of 0.3 Watts. If one could be placed on the end of a wand and supplied with an adequate power source, then a *Lumos* like white light could be emitted. Providing such a bulb could be supplied with enough electrical energy and made robust enough to be operated at high temperatures, it could provide a crude light on the tip of a wand.

A more effective option would be to use Light Emitting Diodes (LEDs) to provide torch light. Unlike incandescent bulbs, they function through electroluminescence. In their operation, the LEDs also don't give off as much waste heat energy as incandescent light sources, so LEDs are more efficient. As a result, they require smaller power supplies to produce the same strength light and they can be made much smaller.

Lumos Maxima

In *The Prisoner of Azkaban*, Harry Potter uses *Lumos Maxima*, to emit the greatest form of wand light that can be produced. It essentially turns a wand into a flood light.

The brighter the light is, the more photons are being released every second. To release more photons, we would have to increase the surface area that is emitting the photons. This would be like causing the wand to give off light from half of its length instead of just the tip. More surface area means more photons given off, meaning the light is brighter.

With incandescent light sources, we could increase the number of photons being released by raising the temperature, although this would slightly change the frequency of the peak radiation and so the color would change as well. So, ruling out the possibility of the wand being a white hot incandescent light source, we're left with a cooler luminescent light source, such as the LED torch on a cellphone or white LED screen.

If the light from the wand is a result of luminescence, then depending on the exact method, it would require an increase in the rate of light producing reactions. This could be done by increasing the temperature of the reactions, or the voltage supplied, or whatever other process underlies the light production.

So, a wand with a brightly glowing tip could be made in a few ways. However, whatever the exact process of creating the light, it would still depend on electrons being excited to higher energy levels then emitting that energy again as light.

WILL WE EVER SEE AN INVISIBILITY CLOAK?

antastical ideas of invisibility garments go back many years. In medieval Welsh mythology, among the 'Thirteen treasures of the Island of Britain,' there is an object called the Mantle of Arthur in Cornwall, which could make its wearer invisible.

In the Wizarding world, the invisibility cloak that Harry Potter inherits is said to be one of the three legendary Deathly Hallows, assumed to be created by Death himself for one of the Peverell brothers in the 13th century. In the legend, the original intention of the cloak was to allow the wearer to go forth without being followed by Death, but generally it was to hide a person from their enemies. Not surprisingly, in the real world the idea of having a garment or device that can render something relatively impossible to see is very attractive; especially to the military.

Camouflage

For years, the military have made use of camouflage clothing to make it harder for enemies to find them. Often camouflage involves wearing colors that match one's surroundings as well as using shapes and disruptive patterns to break up the perceived outline of the object. It basically messes with the way we perceive what we see rather than making the object completely invisible.

In nature, the ability of an organism to blend into its environment is known as crypsis, and it provides a survival advantage to the organism. For example, due to its color, a polar bear is harder to see in its mostly snowy environment, while a leaf insect has both the color and shape of leaves in the tree that it inhabits. As the adaptive features of these animals do not change with time and place, these animals are using passive camouflage.

On the other hand, chameleons and octopuses use active (also known as adaptive) camouflage as they adapt the color, pattern, texture, or shape of their skin to match whatever surroundings they find themselves in. Octopuses, in particular, can alter all of these features to mimic objects in their surroundings, whereas a chameleon can mainly just change color and pattern.

In the wizarding world, wizards have their own chameleon-like ability. It's called the Disillusionment Charm and it makes the subject's body take on the color and texture of the things behind it. Again, with all of the above examples, the target is still visible but disguised in a way that makes it hard for us to recognize what they are.

Adaptive Camouflage

Adaptive camouflage has been investigated by various groups of muggle scientists, for human as well as vehicle applications. In 2003, a professor at the University of Tokyo developed a system named Optical camouflage, which uses what is described as Retro-reflective Projection Technology (RPT). A camera captures the background that's obscured by an object and then an optical projector displays this image onto the front of the object in real time. The object is covered in a special retroreflective material which acts as a screen for the projection.

In 2012, the UK series *Top Gear* fashioned a system for a Ford Transit. They surrounded the van with four walls of flat screen televisions that faced front, back, left, and right. Opposite each wall of televisions, they positioned cameras that relayed live images of the view on the other side of the van. Mercedes-Benz used a similar setup in an advert for their F-CELL Hydrogen fuel technology. Rather than using flat screen TVs for displays, they carpeted one side of the car with arrays of LEDs.

More recently, in the US, a company called Folium Optics has been working on a technology that could be used on combat vehicles. The tech uses an array of hexagonal cells that can switch color to match its surroundings. The cells are also reflective so the system doesn't require much power while naturally matching the brightness of the ambient lighting conditions.

Despite best efforts a downside with camouflage is that if the concealed object moves, or is viewed from different angles it tends to lose its stealthy advantage. A truly invisible object, such as evoked by the cloak of invisibility, shouldn't have this problem. So how could something actually become invisible?

The Eye's the Limit

Invisibility is all in the eye of the beholder, which in this case, is human beings who see things via visible light. The role our eyes play is to make use of light coming from objects in the surroundings. Eyes first evolved underwater, which is a place where only certain light frequencies can penetrate. As such, our eyes became most sensitive to those water penetrating frequencies; specifically, visible light.

Microwave, infrared, and ultraviolet (UV) radiation are absorbed by water molecules meaning they do not penetrate water well, eyes would have no need to evolve sensitivity to those wavelengths of light. Although some insects, like bees, can perceive some frequencies of ultraviolet light.

The reason we see the world at all is because our eyes can absorb the visible light waves coming from objects. The process requires the light to be absorbed in special parts of the eye, particularly in structures called rods and cones, which exist in the retina. We have about 120 million rods and over 6 million cones in our retina. The rods are sensitive to low-level light, while the cones respond to the visible light frequencies associated with colors.

Anything that doesn't give off visible light is effectively invisible to us, but some animals such as snakes are able to detect infrared radiation. Snakes do this via special pit organs on their faces that allow them to detect infrared light up to one meter away. This ability is why Voldemort's snake, Nagini, can see Harry and Hermione while they are under the invisibility cloak. This indicates that Potter's invisibility cloak is only transparent to certain light frequencies and possibly only visible light. Would it be possible to make a visible object invisible to the human eye?

Oh, That Old Trick!

For something to be truly invisible to the naked eye, it has to let light straight through, meaning that the atoms of the object do not noticeably disturb the light waves as they pass through. This is why substances such as water, glass, plastic, and air appear transparent. However, we can still see them. How is that possible?

Other than dirt or smudges on the surface of transparent materials, we can see them based on the way light is affected as it travels through them. For example, as it travels from one transparent substance to another (e.g. air to glass or glass to water) any light that hasn't been reflected or absorbed can change direction, depending on the angle it approaches the interface between the two substances. This particular bending of light is called refraction.

Refraction happens when the substances don't have the same refractive index. The refractive index provides an indication of how much the speed and direction of light can be effected as it travels through transparent materials. When light travels through a glass in air its path gets perturbed because air and glass have different refractive indices. Along with reflec- tion and absorption of light, this refraction affects the original light path, revealing the presence of the glass.

Some materials do have roughly the same refractive index though, for example cooking oil and Pyrex. This means that light is not refracted as it travels from one to the other; making it appear like the light has travelled straight through the oil and glass without disturbance. This is the basis of a science 'magic' trick that causes Pyrex glass to apparently disappear when it's placed into a container filled with cooking oil. In essence, it's the oil that provides the cloak.

This trick only works because the Pyrex glass is already see through and clear, but to make an opaque or tinted object become invisible would require a different technique.

Metamaterials

Scientists have been looking into technologies that can direct light around objects, rather than trying to somehow make the actual object transparent to light. In 2006, physicist John Pendry came up with an idea which was subsequently dubbed the 'invisibility cloak'. The underlying technology manipulates light by making use of metamaterials.

Metamaterials are specially engineered materials that exhibit properties that are beyond those of naturally occurring materials. For example, they can have negative refraction, which is something not found in nature. The first metamaterials only functioned for longer wavelength radiation such as microwaves and radio waves but researchers have been working to extend this range. For an idea, microwaves have wavelengths between 30cm and 1mm, whereas visible light waves are between 400 and 700 nanometers. A nanometer is a million times smaller than a millimeter.

In 2012, researchers at the University of Texas at Austin successfully cloaked an 18-centimeter tube from certain microwave wavelengths. Their cloak works by suppressing the way light is scattered (reflected in multiple directions) by an object. If the light isn't remarkably scattered by the object, then we can't detect the object's effect on the light, effectively rendering it invisible to those wavelengths of light.

The following year, the same researchers used ultra-thin metascreens to produce what they called a mantle cloak, because it has the advantage of being extremely thin (less than 1 mm thick) and flexible. Co-author of the study, Andrea Alu described it like this. "When the scattered fields from the cloak and the object interfere, they cancel each other out, and the overall effect is transparency and invisibility at all angles of observation." So, could this be used to cloak objects in visible wavelengths?

To make objects invisible to our eyes and not just longer wavelengths, we would have to scale everything else down, including the size of the object being cloaked. This is because the object has to be smaller or comparable in size to the wavelength of the light being used to view it. If the object is much larger, the cloak won't work. So instead of an 18-centimeter long cylinder, the object could only be about 1 micrometer long, i.e., a thousand times smaller than a millimeter.

In fact, there are fundamental limits to the size of objects and wavelengths that can be used in metamaterial cloaks, although different types of metamaterials such as active metamaterials may provide us with more promising possibilities for the future.

Invisibility Cloaks

Adaptive camouflage is a long way from wearable tech, but it's showing promise with bigger objects such as vehicles. The work being done is slowly leading to better techniques for hiding objects from view. Bae Systems have already demonstrated an infrared cloaking system for tanks.

Although we will likely never find a way to make opaque objects invisible to visible light, it is possible to bend the light around an object in a way that provides a similar effect. These invisibility cloaks do exist and are made possible by using metamaterials. However, there are limitations with the maximum size of object that can be cloaked from visible frequencies.

At the moment, no such luck on getting an invisibility cloak to throw over your shoulders, but it's definitely a possibility in particular wavelengths and for objects of limited size.